Susan Barr, David Newman
and Greg Nesteroff

ERNEST MANSFIELD
(1862–1924)

"Gold – or I'm a Dutchman"

akademika"
publishing

© Akademika Publishing, Trondheim 2012

ISBN 978-82-321-0020-0

This publication may not be reproduced, stored in a retrieval system or transmitted in any form or by any means; electronic, electrostatic, magnetic tape, mechanical, photocopying, recording or otherwise, without permission.

Layout: Type-It AS
Cover Layout: Mari Røstvold, Akademika Publishing
Paper: Arctic Volume 115 g
Printed and binded by: 07 Gruppen AS

This book has been published with funding from Svalbards miljøvernfond (Svalbard Environmental Fund)

We use only environmentally certified printing houses.

Akademika Publishing
NO–7005 Trondheim, Norway
Tel.: + 47 73 59 32 10
Email: forlag@akademika.no
www.akademikaforlag.no

Publishing Editor: laila.andreassen@akademika.no

Contents

Foreword 5

Introduction. 9

Ernest in New Zealand (c. 1878–1897) 11

Ernest in Australia (1897 and 1914). 25

Ernest in Canada (1898–1901) 31

Ernest in England (1904–1924). 55

Ernest in Svalbard (1904–1913). 75

NEC cabins in Svalbard 145

Ernest's literary and musical endeavours 161

Ernest's personality – Was he a dreamer or a swindler? 179

References 187

Foreword

For over 30 years, adventurer, prospector and author Ernest Mansfield circled the globe seeking fortunes from New Zealand to Canada to England to Svalbard. He seldom found them but left behind fascinating tales and a legacy of huts and mining equipment that continue to draw arctic cruise ship passengers and tourists. Mansfield's legacy is curious. In some places, he swept in as a powerful source of investment capital and put his stamp on local industry, only to be swiftly forgotten. In other locales, he is well remembered – both as a hero and a villain. He was charismatic and well liked by associates and employees, but his ventures did not always work out. Moreover, while he was a talented and shameless self-promoter, certain aspects of his personal life remain a mystery, as though deliberately erased. Born in London, Mansfield emigrated to New Zealand when he was 16. He soon became an accomplished musician and composer in Wanganui, playing several instruments, giving banjo lessons and working in a music shop. By the age of 27, he was writing short stories and reporting for a local newspaper. He also edited several specialist magazines. It was probably as a journalist that he became interested in prospecting and the excitement of the gold rushes then prevalent in New Zealand. He became a registered broker of goldmine shares on the Auckland stock market, and his speculative and controversial enterprises were well documented in local newspapers. In his early 30s he travelled across Australia to the gold rush town of Kalgoorlie and again involved himself in prospecting and land speculation. Across the Pacific Ocean, however, a bigger attraction loomed: the Klondike gold rush in Canada's Yukon, and he could not resist. Soon his combined prospecting, speculating and journalistic skills came to the fore, and in Nelson, British Columbia, during the late 1890s he developed and heavily promoted the Camp Mansfield mines in the Kootenay Mountains. His skills peaked when he orchestrated a five-day prison stay in the Nelson jail as a ruse to extract finances from his French investors to pay miners' wages. Just as suddenly as he arrived in Canada, he left, and in 1904 established himself with a new young wife in a tiny village in Essex, UK. There has never been gold in Goldhanger, but that did not inhibit his prospecting adventures. He became friends with the local rector and doctor and together they threw themselves into gold prospecting schemes in largely unexplored Svalbard. Dr John Salter, famed for a diary recording travelling and hunting exploits with friends in high places, dedicated a chapter in his *Reminiscences* to their experiences. The three men worked well together, employing skilled Goldhanger men to build huts in Svalbard. While over-wintering alone in one of these remote cabins, Mansfield wrote his first semi-autobiographical and science-fiction based novel, *Astria – The Ice Maiden*, which contains several amazing scientific predictions. The trio formed the Northern Exploration Company (NEC) in 1910, raising substantial capital on the London stock market to develop their gold, marble, coal, asbestos and zinc claims and mines in Svalbard. Over 50 buildings were erected, with many named after the founder (Camp Mansfield), his family (Camp Zoe – his daughter), his wife, mistresses and major shareholders (Camp Morton – the Earl of Morton) and some of these still stand. His marble quarry is known today – 100 years later – by the imposing name of New London (Ny-London). Many Scottish and Norwegian miners were employed over a 10-year period, and newspaper reports about their achievements went around the world. Mansfield's second book, written in 1913, was also semi-autobiographical, based on his early New Zealand and Australian experiences. It is more philosophical than the first as it opines about prospectors, miners and their difficult relationships with the professions, rich investors, bankers and politicians. Also in that year doubts about the viability of the mines emerged. Combined with the Great War, the 1920s depression and the Treaty giving the sovereignty of the Spitsbergen archipelago – now named Svalbard – to Norway, the company never recovered. In addition to his two books and the NEC publications, Mansfield left behind a wealth of newspaper and magazine articles. These, together with Dr Salter's diary and many never-before-published photographs made available by members of his family and relatives of the Goldhanger men involved in early expeditions, give new insight into this extraordinary man and his Klondike-like activities on three continents. This biography is a collaboration between authors in England, Norway and Canada, who independently became fascinated with Ernest Mansfield before joining forces to tell his story.

Acknowledgements

Alexander Turnbull library, Wellington, NZ – For the photo of Wanganui Garrison Band in the 1880s.

The British Library – For the use of E. Mansfield's 1905 Spitsbergen map.

Marion Everiss – For the use of photos of Camp Mansfield, British Columbia.

Richard Gardner – For the loan and use of material from the two NEC prospectuses.

Goldhanger Parish Church – For access to births and marriage registers.

Goldhanger History Archives – For early postcard street scenes.

Lorna Key – For the use of extracts from "Little Totham: The Story of a Small Village".

Dave King – For the initial loan, and then agreeing to supply Ernest Mansfield's two novels.

Shawn Lamb, Touchstones Nelson archivist – For assistance locating photos of Camp Mansfield.

Rosemary Mann – For the use of photos from Charles Mann's album of the 1906 and 1908 Spitsbergen expeditions.

Ross McNeill – For the use of David Booth's daybook and photos.

The National Library of Australia – For extracts from AU newspapers held in: trove.nla.gov.au

National Library of New Zealand – For extracts from many NZ newspapers held in: paperspast.natlib.govt.nz

SMS = Sysselmannen på Svalbard / The Governor of Svalbard and cultural heritage advisors Hilde Tokle Yri and Siri Hoem – For help with photos of Svalbard cabins.

Graphic designer Marcus Thomassen – For helping to inspire the front cover design.

Terry Turner – For present-day photos of Camp Mansfield and Joker lakes.

Peter A. Watson – For research assistance in many ways.

Professor Chris Wainright – For the use of photos taken at Camp Mansfield, Svalbard in 2011.

Ernest Mansfield. Source: Marble Island prospectus, courtesy of Richard Gardner

Introduction

Ernest Richard Mansfield was born in London in 1862. According to his semi-autographical novel *Astria – The Ice Maiden*, he "had to hustle for my own living ever since I was twelve". Maybe this is so, but it has not been possible to find any more details of his early life. From about the age of 16, however, when it seems he emigrated alone to New Zealand, the amount of information about his life steadily grows. Some facts that one would dearly love to know have unfortunately remained elusive. It would seem that Ernest himself was a little careful about which facts – or were they all *facts*? – he wished to have handed down to posterity. In this account the authors have distinguished between what they have uncovered as facts, possible facts and unknown facts. The story has fascinated us more and more as this amazing man's activities have gradually been revealed in all their colourful diversity.

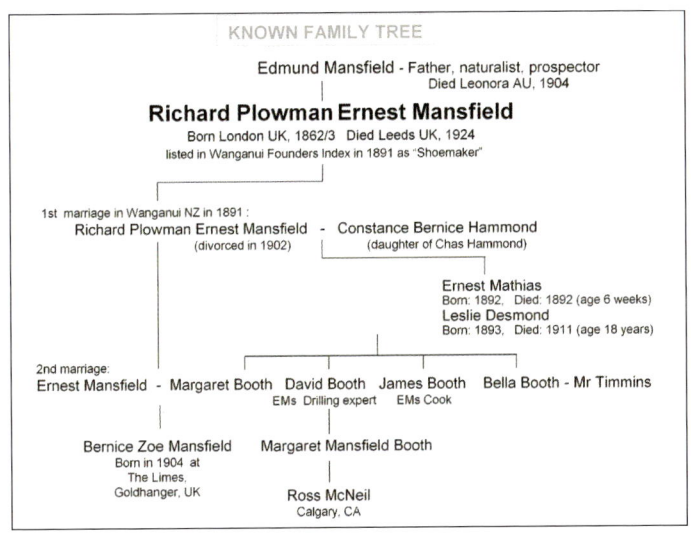

Produced by
David Newman

One of the many abandoned mining attempts Ernest left behind him. Here by the Recherchefjord in Svalbard. Photo: Susan Barr

Ernest in New Zealand
(c. 1878–1897)

In Mansfield's novel *Astria – The Ice Maiden* (1910), the main character states that "before I was sixteen New Zealand was my home" which, if the novel is semi-autobiographical as it is believed to be, would make the year around 1878. The basic facts in the novel have a very clear foundation in Mansfield's own life and it is therefore possible to infer that he in fact did emigrate to New Zealand at the age of 15–16. A "shipping intelligence" report in the *Wanganui Herald* in 1880 lists a "Mansfield," but without a first name one cannot be sure it is the same person.

His presence in New Zealand can be definitely registered from September 1890, when he was 22 years old and he starts to appear in newspapers as Ernest Mansfield in Wanganui (*Wanganui Herald* and *Wanganui Chronicle*). In his second novel, *Ralph Raymond* (1913), the hero is again presented with the statement that "the New Zealand goldfields is where I worked and lived since I was a lad of sixteen and now I am twenty-six" (he was 51 years old when the novel appeared).

Accepting that Ernest probably did emigrate from London at the age of 15–16, the question would be what it was that brought a boy this young to emigrate from England to New Zealand? Nowadays one could imagine that this would be an unusual event, at least if the boy emigrated alone, without his family. In the 1870s this was, however, not unusual. School leaving age was 14–15 and the child was then considered adult. Prospects of jobs and advancement in life were not so good for many in England, and the chances of making it good in the overseas colonies – Australia, Canada, New Zealand – could certainly attract an adventurous and enterprising soul, as Ernest showed himself to be in his later life. In addition the 1870s was a period of organised immigration to New Zealand with 38 000 arriving in 1874 alone (http://www.teara.govt.nz/en/history-of-immigration/8). New

Zealand was at this time in decline, both with regard to wool prices and gold production – the latter being a subject that Ernest was to devote particular energy to some years later. In 1870 the New Zealand government appointed an agent general in London to advertise the advantages of emigration to New Zealand, and in addition free or assisted passages were offered. According to the website *Te Ara (Encyclopædia of New Zealand)* (http://www.teara.govt.nz/en/history-of-immigration/8), almost half of the new immigrants from 1871–1885 arrived on the government-assisted system and three-quarters of these sailed direct from the UK. During the period 1871–80, 100 000 assisted migrants arrived in New Zealand, which was more than half of the total immigration during that decade (http://www.nzhistory.net.nz/culture/immigration/home-away-from-home/summary). The idea was that the influx of Europeans, particularly British, would create an economic boom at the same time as they would take over from the Māoris and expand the ideas of "British civilisation". Germans and Scandinavians were also recruited, but transport was made free for British emigrants from 1873 in order to increase their proportion. At this time, as *Te Ara* relates, there were 53 New Zealand immigration agents in England, 78 in Scotland and 46 in Ireland. The emigrant trickle to New Zealand "became the flood of 1874" (http://www.teara.govt.nz/en/history-of-immigration/8). The new country in the south was, for example, advertised amongst farm workers in England as A GOOD LAND – … A LAND OF OIL, OLIVES AND HONEY; – A LAND WHERE IN THOU MAY'ST EAT BREAD WITHOUT SCARCENESS … (*The Labourers' Union Chronicle*, November 1873).

How and why Ernest ended up in the town of Wanganui is not known. He was perhaps placed there in connection with the assisted immigration scheme. The *Wanganui Founders Index* entry for Ernest held in the Wanganui Museum gives his occupation as "shoemaker" and his father's name as Edmund with an occupation as "naturalist". No "Edmund Mansfield" has been identified in New Zealand records, which is consistent with Ernest recording that he was born in London. This, together with an occupation of shoemaker, would suggest an association with an assisted immigration scheme. There is, however, indication that an Edmund Mansfield who was associated with gold-mining activities in Australia, where Ernest also was involved in short periods, could be Ernest's father (see the chapter *Ernest in Australia*).

Wanganui lies on the River Whanganui in a fertile farming area in the southwest of North Island. It was a major site of pre-European settlement, which was taken over by European settlers under the auspices of the New Zealand Company around 1840. Twenty years later the population was around 2000, with farming as a main objective. The Wanganui website (http://www.wanganui.com/home) is still

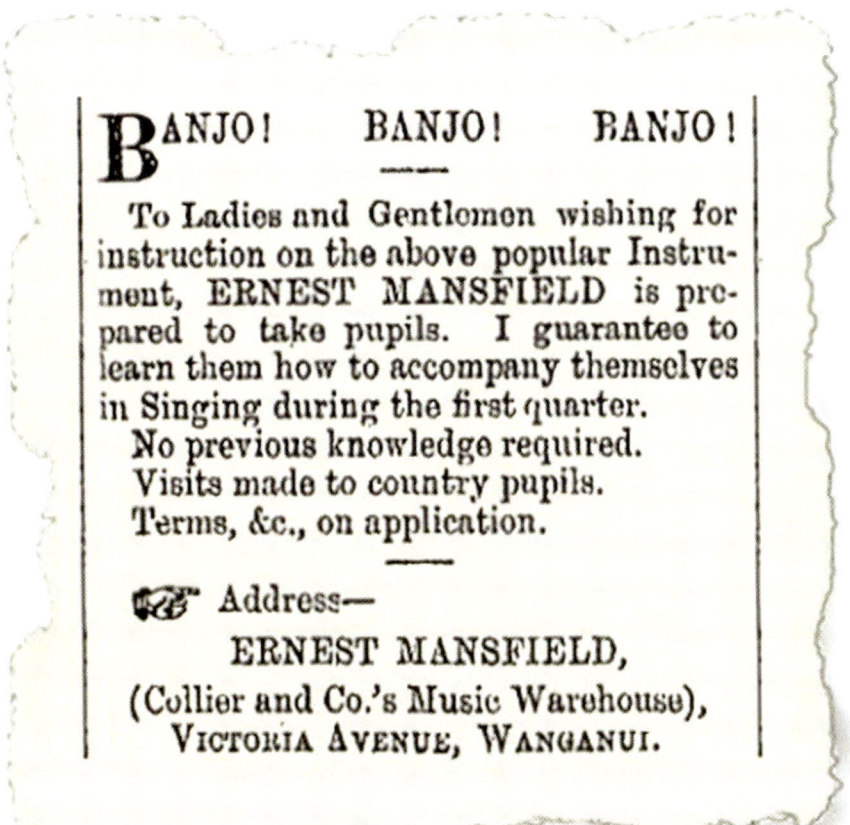

Ernest's early banjo advertisement. Source: Wanganui Herald

today advertising for families and others to move to the area. A town bridge was opened in 1871 and a railway bridge in 1877. By 1886 the town had good railway connections both to New Plymouth further west along the coast and to Wellington in the far south of North Island. The Royal Wanganui Opera House was built in 1900 in commemoration of Queen Victoria's long reign (Wikipedia).

Ernest Mansfield's first appearances in the newspapers in late 1890 are in connection with advertisements for banjo lessons, in which he promises to teach pupils to play and sing. The address is given as Collier and Co. Music Warehouse, Victoria Ave., Wanganui. He claims to have been a pupil of "the celebrated Herbert Ellis, the London banjoist and composer" (*Wanganui Herald* 2.9.1890 and 30.10.1890). At the same time he is writing and publishing short stories. In the

Ernest in New Zealand | 13

Wanganui Herald 31.10.1890 evening supplement, his "original and locally-written story" *A Terrible Ride* is recommended. In fact, a short story with this title was published anonymously a year earlier on 19 October 1889 in the *New Zealand Star*. The note in the *Wanganui Herald* that it is "in that gentleman's best style" also indicated that this was not Mansfield's first published short story. Two weeks later another story appears – *Two Chums* – from an author "whose literary efforts are always perused with interest" (*Wanganui Herald* 13.11.1890). The plot is described as "most romantic and engrossing ... A number of the most interesting scenes being laid in New Zealand, while the characters give excellent scope for some intensely dramatic situations". In January 1891 *How I got my Name* appeared, being described as "a humorous little story" by the *Wanganui Herald* (14/16.1.1891) and "a True Authentic Account of a Christening Party" in the introduction to the story (*Wanganui Herald* 17.1.1891). The story takes place on a train in south England and is, as advertised, just a short and humorous story about christening a baby, with references to the musicality of the baby's cries as being the only apparent tie to Mansfield himself. In June there appears *A Real Romance*, again "an interesting and locally-written story" written by an author who "will be remembered as the author of "Three Christmas Days", and several other excellent literary efforts" (*Wanganui Herald* 6.6.1891). References to six short stories by Mansfield have been identified in contemporary New Zealand newspapers (http://paperspast.natlib.govt.nz) and three were published in full. It is not known whether banjo lessons and story writing provided Mansfield with enough income to live on in this early phase of his life, and he may well have had another occupation such as shoemaker, although there is no evidence of this and it does not seem to quite fit in with his character.

In January 1891 the *Wanganui Herald* reported that a Mr Mansfield sang "There's Gold in the Mountains" at a concert in the Oddfellows Hall. At around the same time in June 1891 "Ernest Mansfield" is recorded as the honorary secretary of the Wanganui Garrison Band in several local newspaper adverts. Reports of a "bandsman Mansfield" associated with this band can then be traced back to a first reference in October 1890. There are many other references to "bandsman Mansfield" and "E. Mansfield" in connection with the Garrison Band throughout the 1890s.

Also in 1891, Richard Plowman Ernest Mansfield married Constance Bernice Hammond (https://www.bdmonline.dia.govt.nz). The middle name "Plowman" used here is interesting as a Plowman was involved with Ernest in Canada (see that chapter). Ernest and Constance Bernice had two sons, both of whom died young. Ernest Mathias, born 25 February 1892 (*Wanganui Herald* 25.2.1892), died at home six weeks later, on 9 April (*Wanganui Chronicle* 11.4.1892). Leslie Desmond was born the following year. In November 1901 the examination results

Wanganui Garrison Band in the 1880s with kind permission of the Alexander Turnbull Library, Wellington, New Zealand ref: F- 11756-1/2. Ernest is believed to be fourth from the left in the middle row

of the Wanganui Infants' School showed that Leslie Mansfield passed the first standard (*Wanganui Herald* 29.11.1901), but he died in 1911, aged 18 (https://www.bdmonline.dia.govt.nz). The Mansfields were divorced in 1902, five years after Mansfield had left New Zealand for Australia and Canada, so Leslie Desmond must have stayed with his mother.

The New Zealand *Evening Post* of 8 December 1906 reports that a decree of 24 October 1902 dissolved the marriage between Richard Plowman Ernest Mansfield and Constance Bernice. On 1 October 1902 Constance Bernice had married Stephen Leonard Parsons, thus committing bigamy, for which reason Parsons in December 1906 applied for nullity of their marriage. As will be seen in the chapter *Ernest in England*, Mansfield had a daughter in 1904 with his second wife and named her Bernice Zoe. It seems strange in the circumstances that Mansfield named her after his first wife.

In August 1892 the *Wanganui Chronicle* informs us that Ernest Mansfield performed *Waltzing Round the Waterbutt* at a concert and also tells of a song that had been written and composed by Mr Ernest Mansfield of Wanganui. The song, entitled *The Wheel's the Life for Me*, is about cycling, with music in a merry style and a pleasing chorus. In fact it is suitable not only for bicyclists, as the newspaper informs its readers, but also for "all who like a good wholesome swing in stirring time" (*Wanganui Chronicle* 23.8.1892). The publisher described the song as a great success. It was dedicated to E.A. Mathiers, who would appear to have been a cycling companion. Mansfield sent the song to "that sterling dramatist and author, G.R. Sims" (*Wanganui Chronicle* 5.1.1893) in England who heartily recommended it. George Robert Sims (1847–1922) was a well-known journalist, poet, dramatist and novelist at that time.

Mansfield took his career in music and literature a step further in January 1893 when he announced a new weekly magazine, *The Stage*, which he was to "conduct" and which would be devoted to "music, sport, and the drama" (*Wanganui Chronicle* 7.1.1893). There would be short and crisp notes on all subjects and special stories and notes for the ladies. Brevity was aimed at, the announcement stated, so the first issue would be only four pages. The enterprise was Mansfield's own private project. We see in this connection Mansfield's tendency to and talent at exaggerated sales descriptions, which were to be characteristic some years later of his mining ventures in Svalbard. It was announced that Mr Mansfield "has written largely for the London papers, has contributed to the leading Australian journals and supplied a great deal of reading matter for the principal papers in both islands of New Zealand". A subscription list of 600 people was apparently already assured, consisting amongst others of "all the leading business people", some of whom had already "donated handsomely" (*Wanganui Chronicle* 9.1.1893). *The Observer* (New Zealand) newspaper was not entirely convinced. In addition to the music, sport and drama, it "also has a funny man who ought to be killed off in the interests of the new venture. His alleged humour is weaker than a circus clown's and more cloying than ginger-beer. Otherwise the little paper is readable enough. Success to it" (Observer 4.3.1893).

In the Wanganui Electoral Roll for 1893 held in the Wanganui District Library, there is the entry: "MANSFIELD, Ernest, Victoria Avenue, Wanganui, journalist, residential". In November of this year, he was acting as travelling representative for a new weekly illustrated journal entitled *Fair Play*, published in Wellington. Here he is described as "formerly of Wanganui" (*Wanganui Chronicle* 7.11.1893), so it may be that he moved to Wellington in late 1893? The *Fair Play* journal itself advertised Mansfield from August to November 1893 (at least) as representing them in the Wellington, Taranaki and Hawke's Bay districts (*Fair Play* e.g. 4.11.1893). All 27 issues of *Fair Play* are available online (http://paperspast.natlib.govt.nz). When Mansfield's name appears, it is only to declare that he was their "travelling representative". Most of the articles do not identify authors, many are "from our correspondent", and although there is an article entitled *Music in Wanganui* which is likely to be by Mansfield, there is nothing specific to connect the article to him. Various short stories and poems were also published, but with no clues as to the names of the authors.

However, Mansfield continues to be mentioned in the Wanganui press. On 1 September 1893, the *Wanganui Chronicle* reported that he was responsible for a new theatrical publication entitled *The Curtain* and in 1894 he was identified as the publisher of the *A1* periodical, which also went under the name of *Gems of thought from noble thinkers*. This was produced in at least 77 editions between 1894 and 1896. On 13 July 1894, the *Wanganui Chronicle* referred to "our new A1 Supplement", and on 25 February 1895 the newspaper reported on the first meeting of A1 Company shareholders, with Ernest Mansfield in the chair.

Fair Play magazine advertisement from 1893. Source: Fair Play magazine.

And still there was more. On 25 April 1895 the *Wanganui Chronicle* announced a new *Museum Gazette* as a "neat little journal by Ernest Mansfield, issued every evening".

However, now came a decisive change of course in Mansfield's life. In July 1895 there was news that gold fever had broken out in New Zealand, first in Auckland a couple of months earlier, and now in Wanganui and nearby Marton. Again Mansfield showed his talent for promotion as the *Observer* wrote (27.7.1895) "if the people of these two progressive towns do not suffer a little from the consequences of the epidemic, it will not be the fault of the glowing expectations recently held out to them in newspaper advertisements and pamphlets. One Ernest Mansfield is the promoter of a mine called the Lydia, stated in the prospectus to be situated at Paeroa". Paeroa lies just south of Auckland and was in an area full of gold mines at the time. The *Observer* reported that "it should bring investors rushing to join in". However, the newspaper was obviously sceptical and ironic about the whole matter, which admittedly glows with the same kind of enthusiasm that reappears in the fabulous reports of marble finds in Svalbard that Mansfield was to send out some years later. The Lydia Gold-mining Co. is in fact made fun of in this newspaper report.

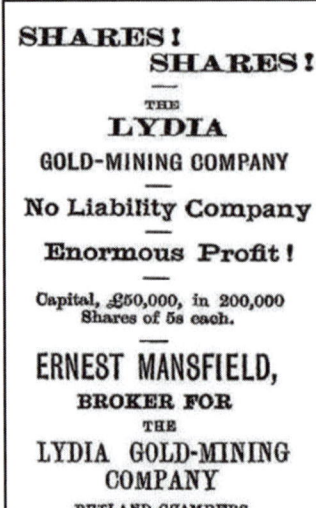

Shares advertisement from 1896. Wanganui Chronicle 1895 & 1896

In the period between July and September 1895, a series of newspaper articles, telegrams and letters were published in the Wanganui press that indicate that Mansfield had joined forces with the local medical practitioner, Dr Hatherley, to pursue their mutual interest in the Lydia Gold-mining Co. Mansfield and Dr Hatherley clearly had other links: draughts (see below), the local Wanganui museum, he was a brother newspaper reporter on the *Wanganui Chronicle*, they were directors of the A1 Company, and they were Freemasons, which Mansfield also had in common with Dr Salter in Goldhanger (see *Ernest in England*).

In July 1895 large adverts for Lydia gold mining shares appeared in the Wanganui press – "applications to Ernest Mansfield". The episode was not without controversy – editorials and letters appeared in the press expressing scepticism about the mine's assay results, and on 8 July 1895 the *Wanganui Chronicle* published a letter questioning the validity of a "No Liability company" share offering. The letter was signed "Cyanide".

However, the *Wanganui Herald* of 27 August 1895 was of the opinion that the shareholders' meeting was perfectly happy with Mansfield's report of the state of the Lydia Company. It stated that the report indeed "should clear the air of many of the ugly rumours that were afloat". Samples from the mine had been assayed with results that were very promising. Again this bears resemblance to events that were to occur in British Columbia and Svalbard years later. A 100-acre property on the Waitekauri goldfields, just east of Paeroa, was claimed by the Company (*Wanganui Herald* and *Wanganui Chronicle* 28.9.1895), and the property was said to be "one of the best in Waitekauri". In October Mansfield was appointed member of the Free Exchange in Auckland and could thus buy and sell shares on the Auckland market (*Wanganui Chronicle* 12.10.1895). He was then advertising in the *Wanganui Chronicle* to buy or sell shares and by January 1896 he was advertising himself as a "Financial and commissioning agent".

Ernest Mansfield was a man of many interests. In May/June 1896 there were long and indignant exchanges in the *Otago Witness* concerning who had the right to call themselves "New Zealand Draughts Champion." Mansfield signed himself "One of the committee and vice-president" of the New Zealand Draughts Association, which crowned a Mr J. A. Boreham during a championship tournament. However, one David Alexander Brodie of Dunedin argued that the title belonged to him. Among his accomplishments, Brodie once played six games simultaneously while blindfolded (http://www.draughtshistory.nl/polish.htm). He did not take part in the Association's tournament, however, belittling it as "more of a local one than otherwise, and few, indeed, if any, of the representative draughts players of the colony took the trouble to journey to Wanganui." (*Otago Witness*, 14.5.1896).

In his reply, Ernest called this "senseless," "tarradiddle" and "deliberately untrue." "Has Mr Brodie ever attended a tournament in New Zealand that could boast of such a representative gathering as the one so successfully carried out here? Has any other draughts player in New Zealand except those attending the fine championship tournament? I don't think so." (*Otago Witness*, 14.5.1896).

Brodie was not chastened. "Pure ignorance alone accounts for Ernest Mansfield's statements", he huffed. Brodie said the Wanganui tournament "fades into insignificance" compared to one at Dunedin in 1890, where he finished third, though he was only 17. "I several months ago made an offer to play any one in New Zealand, and if this Ernest Mansfield or the players he names consider themselves superior to me they are very foolish in not accepting my offer," he added. "When Mr Mansfield has a little more experience and has attended some more tournaments he will cease to take such a narrow view ... I cannot take Ernest Mansfield's words as echoing the sentiment of the association, for in a matter of this kind they would be more careful in the election of a suitable person to defend them." (*Otago Witness*, 28.5.1896).

Despite his exasperation, Ernest extended an offer to Brodie to become part of their organisation, which then numbered close to 200 members. "Mr Brodie, why not bury the hatchet and join us, and let us all work together to send the grand old game along?" he asked (*Otago Witness*, 11.6.1896). There is no record of any further reply.

In September 1896 the *Wanganui Chronicle* reported that E. Mansfield had a role in the local museum, organising their collection of precious stones.

Also in 1896 Mansfield registered the New Munster Gold Mining Company with 1500 shares each held by himself – with occupation "journalist" – and Constance Bernice (*Observer* 31.10.1896). On 12 October 1896 the *Wanganui Chronicle* reported that Mansfield had completed the sale of the Waitekauri Golden Lead mine to a London company with the result that the lucky New Zealand investors would receive a handsome return for their investment. The newspaper stated: "Mr Mansfield formed a syndicate last year to partly purchase and provide the money to develop the Golden Lead property, and the shareholders now will receive a very handsome return for their investment, amounting to £1,562, cash and shares, for

Produced by David Newman

£100 invested… There are Several large shareholders along the coast and Mr Mansfield holds a considerable interest in the property". So the signs are that Mansfield made a considerable amount of money from this deal. This could well have been the Hikutaia Gold Syndicate Ltd. listed in the 1898 "Statement by the Minister of Mines" (see below), which also refers to the Lydia Claim.

At the end of January 1897, Mansfield, "who is well and favourably known on this coast" (*Wanganui Chronicle* 21.1.1897), was to leave "on an extended trip to the Old Country" in connection with mining and other business which were also the concern of "several local residents and prominent settlers". These friends and admirers gave him a handsome send-off with the presentation of a gold chain and pendant inscribed "Presented to Mr E. Mansfield by the Wanganui Gold Syndicate, January 20th 1897". His speech of thanks was met with three cheers. It seems therefore without doubt that Mansfield's mining speculations in New Zealand at this point had paid off. The *Wanganui Chronicle* reported that some 170 members of the Cosmopolitan Club had attended the send-off, with Mansfield's energetic endeavours to find gold and his "extensive northern mining transactions" being praised in addition to his "well-known generosity". The hope was expressed that his efforts, "both locally and in connection with the Thames" would produce a substantial profit for all concerned (*Wanganui Chronicle* 22.1.1897). The *Wanganui Herald* followed his progress, reporting on the same day (22.1.1897) that E. Mansfield and Dr Hatherley had left Wanganui by train for Auckland (two separate reports in the same paper). However, in May 1898 the *Thames Star* carried an advertisement placed by the Warden's Court stating that … "Ernest Mansfield has failed to pay rent on claim and cannot be found", which would indicate he had left the area without leaving any forwarding address.

In the document entitled "Statement by the Minister of Mines" dated 1898, there is a list of mine registrations, but no "Lydia Gold-mining Company" is identified, and the only "Lydia mine" is shown as owned by the Hikutaia Gold Syndicate Ltd. Various reports about the mine in the 1898, 1903 and 1907 versions of the document indicate that it was never profitable.

A note in the *Evening Post* (New Zealand? 13.2.1899) dated London, December 30 informed that Mr Ernest Mansfield, "who will be remembered in New Zealand, particularly by draught players, among whom he took a high place", was apparently in northwest Canada and representing a French syndicate with some very good mining claims. In June the *Wanganui Chronicle* (8.6.1899) could inform its readers that Mansfield was still in British Columbia (passed through Vancouver) and the following day had a long article about Mansfield – "one of the most successful canvassers which this district has known" under the heading *Mr Ernest Mansfield in British Columbia*. The article reminded readers that Mansfield "like most men, had his peculiarities", but certainly had a strong belief in his ability to succeed in whatever he tried. He had ended his stay in New Zealand with large-scale mining speculations and was in London on the point of placing the New Zealand gold claims on the market when the Klondike boom knocked the market out of all gold speculations outside of British Columbia, whither he immediately had hastened. The article ended by stating that Mansfield hoped to spend his next Christmas in Wanganui (*Wanganui Chronicle* 9.6.1899).

The *Wanganui Chronicle* kept its readers informed about Mansfield's activities in British Columbia, including various business travels to Europe in between, through articles in Canadian newspapers. For example, an extra long article on 22 December 1899 ended by informing Mansfield's Wanganui friends that he would be back there on 17 January – having obviously not made it for Christmas. British Columbia newspapers continued the information about Mansfield's whereabouts by reporting that he was to sail for New Zealand on 18 January 1900 on "a business and pleasure trip of a few months" and was expected to return around 10 April (*Silvertonian* 6.1.1900). The article went on to state that Mansfield had earned over the past year an enviable reputation amongst his prospector business associates and that all his friends in Slocan (near to Nelson) wished him bon voyage and an early return.

Other New Zealand newspapers also kept their readers informed about Mansfield's activities in Europe after he transferred from British Columbia to England around 1904. For example in November 1906, there was a short article in the *Wellington Evening Post* reporting that "Mr Mansfield of Auckland had returned to London having been coal-prospecting in Spitsbergen". On 24 November 1910 the *Wanganui Chronicle* printed a long article, taken from *The Times* of London, under the heading of *An Arctic Adventure*, which presented Mansfield as the hero of a shipping incident off Spitsbergen when he rescued some Norwegian explorers and sailors and saved their ship (see *Ernest in Svalbard*).

Two local newspapers, the *Wanganui Chronicle* of 15 September and *Grey River Argus* of 27 September 1911, both ran the long stories about Mansfield's early exploits on Spitsbergen (see *Ernest in Svalbard*). The former article was entitled *Sands of Gold* and the latter *In Search of Gold*. Both recorded the meeting between Ernest Mansfield, the Rev. Gardiner and Dr Salter that is the commencement of the activities Ernest was involved with in Svalbard. On 31 July 1912 an editorial appeared in the New Zealand *Thames Star* reporting the receipt of a "telegraphic message" from one Ernest Mansfield in Spitzbergen to friends in New Zealand after a 15 year absence. The article summarised his history and described him as "a wonderful salesman and canvasser" who has, "it is said, been in tuch [sic] with the world's leading financiers, and his headquarters have alternated between London and Paris".

In June 1913, at the time of the publication of Mansfield's second book *Ralph Raymond* in the UK, a very long article and review was published in the *Wanganui Chronicle* entitled *Ernest Mansfield – Explorer, Prospector, Promoter and Author*. This was some 14 years after Mansfield had left New Zealand. The article began with a resume of his career in New Zealand, Canada and the UK. It then briefly summarised the plot of the novel, but the majority of the review concentrates on the author's philosophical views expressed in the book rather than the murder/mystery plot. The review had two sub-headings: *There is no Death* and *The Prospector & Promoter - How the Sharks Work* (see the chapter *Ernest's Literary Endeavours*).

On 1 November 1913 a single advertisement for Mansfield's *Ralph Raymond* appeared in the *Wanganui Chronicle* referring to Mansfield as "Erstwhile from Wanganui".

The last article identified in New Zealand newspapers which refers to Ernest Mansfield was on 20 October 1915 reporting on a visit by Dr Hatherley to the UK. He wrote: "Mr Ernest Mansfield, an erstwhile Wanganui resident, is fighting for King and country". It is not known to what this referred.

RAYMOND FLUNG HIMSELF AT THE EXHAUSTED AND BATTERED MAN

Page 257

Ralph Raymond's heroic activities in Australasia were designed to convey a public image of Mansfield himself. Illustration from Mansfield's second book, in the current authors' possession

Ernest in Australia
(1897 and 1914)

Ernest visited and conducted business in Western Australia on two occasions, which were separated by a period of eighteen years. It is known from reports in Wanganui newspapers that he left New Zealand in January 1897, and there is evidence of him visiting the Kalgoorlie area in Australia in 1897 and then again in 1914. Other than that they were related to gold prospecting, little is known about his Australian-based activities; however several short reports relating to his visits are presented here in chronological order.

In May 1896 the *Kalgoorlie Western Argus* (28.5.1896) reported:

> Mr S Mansfield representing the Good Luck Syndicate of London, formed chiefly by South African millionaires, has arrived here. He is looking for properties to invest in, also city lands, and will probably make Kalgoorlie his headquarters.

Although the timing, location and context would suggest that this could have been Ernest, the article refers to "S Mansfield", and Ernest was believed to have been in New Zealand at this time. It could perhaps have been a relative.

In September 1897 in a long article, the *Wanganui Chronicle* (17.9.1897) reported that the New Zealand representative at Queen Victoria's diamond jubilee celebration in The Strand, London, was Mr Ernest Mansfield. The event was organised by the Hurst family for the benefit of "Australasians". Mr G. P. Hurst had "traversed the waterless wastes of central Australia which surround Coolgardie". Mansfield clearly took a prominent role in the event and gave a speech thanking his hosts for their hospitality and his "deep sense of gratitude for the many kindnesses he had received from the Hurst brothers and their mother". The guests retired to the Hurst family home in south London which was called "Kalgoorlie Lodge". There is evidence of Good Luck Syndicates being involved in gold mining in Kalgoorlie in Western Australia and in Africa and India at the turn of the century. In a lengthy article in the *Wanganui Chronicle* (9.6.1899) relating to Mansfield's activities it noted: "Mr Mansfield…has been in Africa, Australia, and New Zealand". It is also known from Western Australia trade directories that the Hurst family operated goldmines under other names in the Kalgoorlie region.

In April 1903 the *Wanganui Herald* (23.4.1903) carried a long article entitled: "Western Australia by a Young Wanganui boy" in which the author describes life in the Kalgoorlie region goldfields. The article is not attributed to Mansfield, but from what we know of his connection with the Wanganui press at that time it is highly likely to have been written by him – even though it would have been pushing facts to describe himself as "young". He was 40 years old by this time.

In November 1904 the *Adelaide Advertiser* (15.11.1904) reported the suicide in Leonora of "a well known prospector E. Mansfield" and a month later the *Kalgoorlie Western Argus* (20.12.1904) reported the death and "intestate estate" of a Mr Edmund Mansfield of Leonora, near Kalgoorlie in Western Australia. Back in 1891 while living at Wanganui, Ernest had named his father as "Edmund Mansfield". One wonders if it is too much of a coincidence that a person of this name lived in the very same small mining town in Western Australia that Ernest returned to on at least two occasions.

Mansfield wrote in his first semi-autobiographical novel *Astria - The Ice Maiden* (Mansfield 1910):

> Before I was sixteen New Zealand was my home, and I saw a good bit of that country, for I was always on the go. Then the Coolgardie rush broke out, and off I was like a shot. It was no great shakes, that trip, I can tell you. Three hundred miles tramp, and dirty water, when you got it – which sometimes you didn't! And the sun ! – my word, it blistered you, and the country you had to go over was as dry as a wooden god! Then I went all through the Australia that was before the Commonwealth days, and was there when the great Klondyke broke out, and off I was again!

His second semi-autobiographical novel *Ralph Raymond* (Mansfield 1913) includes many references to Australia. In Chapter 1 he wrote:

> As to your request for a reference, I can only give the names of my mates on the Australian and New Zealand goldfields. That is where I have worked and lived since I was a lad of sixteen and now I am twenty six.

and later in the chapter he writes:

> This is my daughter Berice, and this gentleman is Mr Ralph Raymond, who sails in the Maharajah as discoverer and manager of the Good-luck gold mine which your mother, myself, Mr Raymond and you are all equally interested.

As Mansfield's books were written in 1910 and 1913, the references to Australia must be to the earlier 1897 visit.

In the *Northern Exploration Company Prospectus* (NEC 1911), there is a short "brief memoir" of Ernest which reads:

> An Englishman by birth. At the age of 16 he went out to New Zealand, where, after a period of engineering work, he migrated to the Goldfields on the West Coast, and there gained invaluable experience as a gold-digger and prospector. Subsequently, he visited Brunnerton, N.Z, where he gathered useful knowledge in coal-mining.
>
> Leaving there he returned to engage in more serious prospecting, choosing the Hauraki and Coolgardie Goldfields. There he secured mining claims which were subsequently very successfully dealt with by Syndicates. In 1897, when the discovery of the Klondyke Goldfields was made, he was engaged by a Syndicate, with a capital of £2,000, to go out there.

The reference to being in Klondyke in 1897 means that his first stay in Australia could not have been longer than a year.

In September 1911 the Perth-based *Western Mail* (16.9.1911) printed an article entitled "Gold Discovery in the Arctic", which was about Mansfield's discoveries in Svalbard and was very similar to other articles published around the world at that time. Then in July 1913 several reviews of *Ralph Raymond* appeared in the Australian press: The *Western Mail* (25.7.1913), *The West Australian* (31.7. 1913) and the *Brisbane Courier* (20.8.1913). All of these articles printed at a time when Mansfield was travelling between England and Svalbard suggest that he was maintaining links with past associates, and perhaps journalist colleagues, in Australia.

In relation to his second visit to Australia, the following article appeared in the *Kalgoorlie Western Argus* (20.1.1914):

> Mr. Ernest Mansfield has been on a visit to the gold fields during this week. He hails from far away, and a much colder clime than ours. He comes from Spitzbergen, the land of the midnight sun, where he is engaged in commercial enterprises on behalf of a big English company. He is a fine raconteur, and tells thrilling stories of his experiences amid the ice and the snow. That is refreshing to listen to, especially with the temperature away up in the hundreds. Mr. Mansfield is the author of two books, entitled "Astria" and "Ralph Raymond."

In February 1914 the *The Kalgoorlie Western Argus* (24.2.1914) reported on a Warden's Court case in Leonora, Western Australia:

> ... the Harbour Lights mine was at present under option to an English company represented by Mr E Mansfield.

and the *Sunday Times* of Perth (8.3.1914) reported:

> The Great Unknown at Marda has been placed under a three month sampling option to Mr E Mansfield, acting on behalf of London investors. At the end of the sampling period and all things being satisfactory a deposit of £3000 will have to be paid.

In October 1915 the *Wanganui Chronicle* (20.10.1915) reported that Mansfield's former friend and associate in New Zealand, Dr Hatherley, stated on a visit to London that Mr Ernest Mansfield was "fighting for King and country", so presumably he was in Europe at this time and not in Australia.

In an obituary of Ernest written by Birger Jacobsen (*Tidens Tegn* 16.1.1925) he wrote:

> He was equipped for his work, with internationally rich experience as a prospector in Australia, Africa, British Columbia and Alaska.

And in a second obituary, Carl S. Sæther (*Tromsø* x.1.1925) wrote:

> He has earned and lost the wealth of Australia, the Klondyke, in Africa, and worked elsewhere in the world around.

Finally, Dr Salter's diary published after his death (Salter 1933) included this sentence about Mansfield:

> He went right across Australia in the dry season, a thing that had never been done previously. What was more, he was accompanied by another man whom he carried on his back during the last three days.

With regard to the Australian locations: Kalgoorlie City is situated in an isolated part of Western Australia 600 km east of Perth. Coolgardie is 50 km to the southwest of Kalgoorlie and Leonora is 200 km to the north. The three locations are well known for their associations with gold mining. The Perth to Kalgoorlie railway opened in 1896, but the completion of a rail link all across Australia took many more decades. The Harbour Lights gold mine is located 3 km north-west of Leonora and is still operating. The Marda goldmine is in a remote location 200 km north west of Coolgardie and is also still in operation.

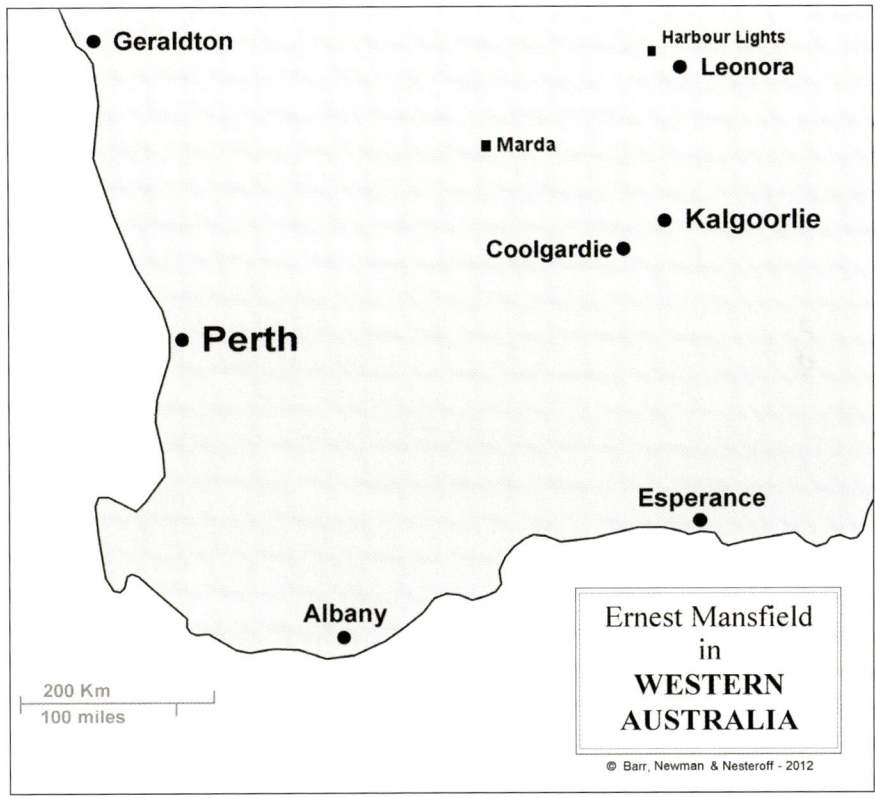

Produced by David Newman

Mansfield Creek, British Columbia. Postcard from the 1890s in the current authors' possession

Ernest in Canada
(1898–1901)

On arrival in England, Ernest placed several gold properties on the London market for a New Zealand syndicate and was reportedly about to close a big deal when word came of a massive gold rush in Canada's Yukon territory (*Wanganui Chronicle*, 9.6.1899).

Financiers suddenly turned their attention to the Klondike boom, and Ernest was forced to do the same. Not that he likely needed much encouragement since he was hardly immune to this siren song. He travelled to the Yukon for a first-hand look, arriving in April 1898, but his stay was transitory and left little trace (*Wanganui Chronicle*, 9.6.1899 and Jacobsen 16.1.1925). In Birger Jacobsen's obituary of Mansfield, it is suggested that he spent time in Alaska as well, but evidence of this has not been found. By summer, he migrated to the West Kootenay district of south-eastern British Columbia, which had experienced its own boom.

Ernest in West Kootenay

Only a decade earlier, this part of the province was seldom seen by non-aboriginals; it was too inaccessible for its mineral wealth to be profitably removed. Things began to change with the discovery of massive ore deposits near what became the cities of Nelson and Rossland. Within a few years, rail lines were built through rugged, heavily timbered country, and the Kootenay River was harnessed to provide power to the burgeoning mines. By the time Ernest arrived, Nelson, on the river's shore, was recently incorporated, and shedding its wild-west flavour in favour of cosmopolitan trappings. Grand sternwheelers plied the river and lake, ferrying passengers and ore.

Representing the Klondyke Champs d'Ore Syndicate Ltd. with £2000 in hand, Ernest spent two months scouting properties.

This British company, employing French capital, evidently had great faith in Ernest for it paid him £1000 per year in salary, travel and general expenses, plus a quarter interest in all properties taken up, and a month's annual vacation with £100 for additional holiday travel (*Nelson Daily Miner*, 25.10.1900). In October, he visited the Joker and Derby claims, staked at the head of the south fork of the Kaslo River by G. W. Taylor, A. G. Lambert and Robert McGregor. Although silver and lead were the region's common currency, these were gold prospects. They were also at extremely high altitude – the highest, in fact, in the region – and now inaccessible due to sudden, heavy snow *(Wanganui Chronicle*, 22.12.1899). Undeterred, Ernest set off from another direction to reach them – and nearly died in a raging blizzard on a glacier, whose wide crevices could only be seen from a few feet away. He "returned to his mining camp long after the other members of the party had quite decided they were lost." Having inadvertently become one of the first men to cross the glacier, he named it after Lord Kitchener (Beaton 1.1902). According to *The Kootenaian* (30.8.1900): "The Kitchener glacier ... was named by Mr. Mansfield at the time Lord Kitchener was making his machine campaign against the Mahdi ..." The glacier was previously and subsequently called Kokanee, from the Sinixt native people's word for the region's once-plentiful salmon.

Practically no work had been done on the Joker, but Ernest was sufficiently impressed to buy the claims. He cabled the syndicate, which accepted his recommendation and increased its capital to £10 000 so development could begin. Despite the difficulties of transportation and impending onset of winter, Ernest rushed in supplies and men, built cabins and started the first vertical tunnel. His optimism was rewarded, for with every foot of depth, the ore increased in richness, and by season's end, the shaft had reached 75 feet (*Wanganui Chronicle*, 22.12.1899 and *Nelson Daily Miner*, 19.10.1900). Various assays put the value of the richest ore at $60 to $70 per ton, and the average value at $40 (*Kootenay Mining Standard*, 7.1899, p. 58)

Ernest went to London in February 1899 to secure capital, and within ten days founded a company – Excelsior Gold Mines of British Columbia Ltd. – and raised £40 000 from various investors. This was the beginning of a tangled corporate web. He may not have been joking when he remarked "I never visit London without forming one or two syndicates or companies" (*Wanganui Chronicle*, 9.6.1899 and *Nelson Daily Miner*, 19.10.1900). According to the *British Columbia Gazetteer and Directory of Mining Companies*, (1900–01, p. 61), Excelsior Gold was licensed in B.C. on 8 Aug 1899 with £200 000 in capital and had its head office at 58 Gracechurch St. in London. Jules Justin Fleutot was managing director and attorney, G. Lawbere was secretary and Ernest was a director. Klondyke Champs d'Ore was the vendor for Excelsior Gold Mines, Ernest explained. The subscribers who came up with the initial working capital were advised to take common rather than preferred shares[1], but they refused, causing huge headaches later *(Nelson Daily Miner*, 20.10.1900).

Returning to Canada, Ernest was joined by J. J. Fleutot, a French electrical engineer who worked on the Trans Siberia rail line, and was offered a high-ranking position by the Russian government. He acted as Excelsior's managing director and attorney and would also oversee erection of machinery to treat the ore from the Joker *(Nelson Daily Miner*, 20.10.1900, *Kootenay Mining Standard*, 7.1899 and Porter, 2006). They arrived at the claims in April, and Ernest pronounced himself "thoroughly satisfied" with the work to date *(Nelson Daily Miner*, 9.4.1899). The problem remained, however, with hauling the ore out.

At a public meeting convened by the mayor of Kaslo, the town nearest their claims, Mansfield and Fleutot asked locals to agitate for a new government wagon road to their properties, for the existing road was in deplorable shape. The presentation was warmly received, and a delegation was appointed to make the case for its immediate construction (*Wanganui Chronicle*, 9.6.1899). As one local newspaper wrote in boosting the proposal, "The development of the properties on the south fork must bring trade to Kaslo. For 20 miles from here in that direction there are claims on which large amounts of money have been spent, and the owners are now waiting for a wagon road to turn them into shipping propositions" *(Wanganui Chronicle*, 9.6.1899).

Mansfield and Fleutot travelled to Victoria, British Columbia's capital, to further press their argument. They were received favourably there as well. En route, Mansfield told a reporter in Vancouver that they planned to install a 20-stamp mill and hydropower plant that would likely consume half their capital *(Wanganui Chronicle*, 9.6.1899). He gushed that he was "a great believer in British Columbia. He has been in Africa, Australia, and New Zealand, and was connected with several properties, but the Kootenay is a much finer country, he thinks" *(Wanganui Chronicle*, 9.6.1899). Ernest also encountered a Wanganui acquaintance, furniture salesman John Anderson, who had settled in British Columbia. "He was quite elated over the chances held out to the energetic in this part of the world," Ernest said, "and expresses himself thusly: 'I reckon a man's a fool if he don't make money in British Columbia.'" (*Wanganui Chronicle*, 8.6.1899).

Despite the delegation and Ernest's optimism, the government said no money could be pledged toward the wagon road "but that they would, if the road was built, see that an appropriation was made next year or grant the Excelsior Mining Company a charter to collect toll to compensate them for their expenditure." (*The Koote-*

---

1 Preferred stockholders receive dividends before common stockholders. If a company is liquidated, they are also paid out first. Typically, preferred stockholders receive regularly scheduled dividends, whereas common stockholders are paid dividends at the discretion of a company's board of directors.

naian, 11.5.1899). Ernest and Fleutot resolved to build a trail which could later be widened into a road. Starting at the Montezuma mine, it would be 12 miles to their proposed mill site, and another four to the Joker (*The Kootenaian*, 11.5.1899). Fleutot offered, however, that if anyone would provide half the cash needed, he would build the road as soon as weather permitted. It was estimated the total cost would be $8000, but no one came forward with the funds (*The Kootenaian*, 25.5.1899.)

Ernest was soon on the move again, heading to Vancouver, then London, Paris and Lille, France (*The Kootenaian*, 1.6.1899 and *Wanganui Chronicle*, 4.8. 1899). In his absence, a gang of men began work on the trail, although Excelsior continued to complain bitterly of the existing road, "that one of the bridges is gone and that the fixing of the existing trail is almost as much work as making a new one". They reached the Joker within two months (*The Kootenaian*, 8.6.1899, 6.7.1899 and 3.8.1899). Excelsior also doubled its capital to £360 000 (*Nelson Daily Miner*, 19.10.1900). By the end of August 1899, four buildings were complete and a shaft house was nearly finished. Fifteen miners were hired and paid $3.50 each per day. The company now owned eight horses and also hired packers to bring 15 tons of supplies up to the mine (*The Kootenaian*, 31.8.1899 and The *Ledge*, 9.11.1899).

Ernest returned to British Columbia around late September, now acting for a London banker named Rene Laudi. On his behalf, he secured all the ground adjacent to the Joker, amounting to over 550 acres.

Over the next few months Ernest:

- Bonded[2] the Apex group from W. E. Boie of Slocan City, including the Twin Lakes, Green Lakes, Crescent, and Apex claims, on which considerable work had already been done, for $30 000 (*The Kootenaian*, 5.–12.11.1899 and *The Tribune*, 2.12.1899). He also bought the Charleton from Boie and hired him as foreman of his operations (*The Kootenaian*, 12.11.1899).
- Bonded the Green Lake fraction for $4500 from Burt Pearson of Slocan City (*The Kootenaian*, 12.11.1899).
- Noticed some land next to the Joker had not been staked, so promptly did so himself and called it the Philomene (*The Kootenaian*, 12.11.1899).
- Bonded the Champion and Lost Boy from Joseph Carton and Gus Schilling of Kaslo for $5000 (*The Kootenaian*, 30.11.1899). These claims and four others were soon after acquired by Pactolus Gold Mines Ltd., whose president was Nelson mayor H. G. Neelands, and secretary was *Nelson Miner*

---

2   Bonding a mining claim was the equivalent of taking an option to buy. The prospective purchaser made a down payment with the balance due by a certain date. In the meantime, the buyer could assess the property and decide whether they wished to buy it outright. If not, the bond would be forfeited and the original owner was free to sell to another party.

reporter Welford W. Beaton (*Nelson Daily Miner*, 13.12.1899 and *British Columbia Gazetteer and Directory of Mining Companies,* 1900–01). Pactolus also had on its board A. Macdonald, former mayor of Winnipeg and president of Great West Life insurance.
- Bonded the Tony and Glacier for $5000 on a one-year agreement from Walter Clough, Robert Bradshaw and a Mr Long of Slocan City. Development went so well that Ernest bought the properties outright before the deadline (*Spokane Chronicle*, 19.1.1900).
- Bonded the Monadnock from W.H. Crawford for $7000 (*Spokane Chronicle*, 19.1.1900).
- Bonded the Black Hawk and Daisy group on Ten Mile Creek from James Rae and Duncan Graham of Slocan City for $8000 (*Victoria Daily Colonist*, 5.12.1899 and *Slocan Drill*, 11.4.1900).
- Bonded the John A. and Treadwell from Charles Plowman for $30 000 (*The Tribune*, 2.12.1899 and *Nelson Daily Miner*, 19.10.1900), through which the Joker lead ran, as well as the Marguerite, Alice fractional and Bertha. Within six weeks, Ernest exercised his option and paid out the property in full, earning a discount for his backers (*Nelson Daily Miner*, 2.12.1899).

The various claims, now collectively known as Camp Mansfield, were covered in six inches of snow, but this did not interfere with mining operations, which continued day and night in three eight-hour shifts. An assay plant was ordered, and

Source: Wadd Bros. of Nelson BC, courtesy of Marion Everiss

INTERIOR BLACKSMITH SHOP, BERTHA CAMP, CAMP MANSFIELD.

Ernest on the left. Source: Wadd Bros. of Nelson BC, courtesy of Marion Everiss

H. L. McCain hired as assayer. Provincial land surveyor Charles Moore began surveying the Laudi properties (*The Kootenaian*, 12-26.10.1899). Ernest predicted his camp would be "one of the liveliest in the province next year. There will be at least 100 men working there during the winter and fully three times that number next summer" (*The Kootenaian*, 12-26.10.1899).

Three more cabins were being erected to house a large staff over the winter and Ernest brought in "sufficient supplies to last 12 men for a year" via Slocan Lake. Welford Beaton and a pack train of 15 horses accompanied him. Sixty more loads were expected (*The Kootenaian*, 5, 12 & 26.10.1899). In addition to the usual miners' supplies, he brought something unusual: "There will be music in the air at Camp Mansfield the coming winter judging from the number of instruments forwarded to that place last week. In the consignment were guitars, banjos, concertinas, accordions, etc. Evidently the mine workers propose to have harmony in their camp during the several months they will be practically snow-bound" (*The Kootenaian*, 9.11.1899). Did Ernest give high-altitude banjo lessons?

Ernest was well liked and respected by his workers. When the British Columbia government passed legislation in 1899 limiting the work day for miners to eight hours (versus the 10, 12 or longer they had been used to), many owners responded by cutting daily pay from $3.50 to $3, precipitating a nearly year-long strike. Ernest, however, was not on side with those owners, and continued to pay the full wage. "It is a pleasant contrast to note the difference in the actions and speech of such men as Ernest Mansfield, and the chronic non-working agitators opposing the eight hour law," one local newspaper wrote. "Mr. Mansfield is a man of knowledge and foresight. He has millions of dollars back of him, and must know how things are going. He is asking no questions, but when he sees a property that pleases him he takes it up and puts men to work. His time is too valuable to be spent in discussing the property of this law or that; like a man of business he accepts the conditions confronting him and makes the best of it" (*The Ledge*, 18.1.1900). In another telling example, a former Camp Mansfield cook got a job at a mine near Slocan City but was fired before he started. The manager remarked that he didn't want anyone from Camp Mansfield, "as he did not feed his men canned turkey and other toothsome grub" (*Nelson Daily Miner*, 26.10.1900).

In all his time in Canada, only one negative comment appears to have been written about Ernest, when a New Denver man called him "a fake" (*The Ledge*, 12.4.1900).

In December, Ernest rode down to Kaslo in five hours. He was still trying to promote construction of a road at least as far as the Joker mill site (*The Kootenaian*, 7.12.1899). To light a fire under locals, he mentioned that both Slocan City and Sandon were considering building roads to his camp, although those routes

Source: Wadd Bros. of Nelson BC, courtesy of Marion Everiss

Camp Mansfield. Ernest in the doorway. Source: Wadd Bros. of Nelson BC, courtesy of Marion Everiss

would be much more difficult and less desirable. He further noted Canadian Pacific Railway surveyors had been in the area, so lucrative were the ore bodies (*The Kootenaian*, 7.12.1899). The *Nelson Daily Miner* (6.12.1899) said: "The fact that Mr. C. E. Perry of the CPR has been running some preliminary surveys up 10 Mile Creek to Kitchener Glacier in Camp Mansfield has reached the ears of the Kaslo and Slocan people." Ernest estimated that he alone had taken up 50 tons of freight at five cents a pound, intimating that Kaslo merchants stood to lose greatly if the road was not built from their end (*Nelson Daily Miner*, 6.12.1899). At the same time, Ernest closed a $125 000 deal to personally buy three groups of gold, silver, and copper claims from one of the few women involved in mining in those days.

Jennie E. Harris had arrived a few years earlier with her husband Tom, who staked a series of promising copper prospects in the White Grouse and Goat River districts southeast of Kaslo on what became known as Harris Mountain. Backed by his wife, whom a mining journal called "a shrewd Canadian lady", Tom completed the annual assessment work although the claims numbered a dozen and consisted of three groups (Kemp, 5.1900). A few years later, Tom's brother struck it rich with the sale of the LeRoi mine at nearby Rossland and offered him a lucrative position. Although they did not divorce, Tom and Jennie never lived together again. Perhaps as unofficial alimony, he deeded his claims on Harris Mountain to her (Kemp,

5.1900). As president of the Kaslo-Slocan Development Co. Ltd., Jennie appears to have been the only woman listed among thousands of men in a 1901 list of British Columbia mining companies (*British Columbia Gazetteer and Directory of Mining Companies*, 1900-01).

She entertained offers for the claims and had no shortage of prospective buyers. An agent for Standard Oil and a representative of a British company both showed interest, but she accepted Ernest's proposal "as she admires his push and energy and realized that all his efforts in this district had been crowned with success" (*The Kootenaian*, 7.12.1899).

> Mr. Mansfield, who is cautious in his purchases, has given these properties a thorough examination and test, and is sanguine ... that he has a great bargain. Mrs. Harris was more favorably impressed with the record he had made in West Kootenay than any other applicant for her holdings, and she preferred he would be the purchaser ... As soon as possible, with his characteristic go-aheaditiveness, Mr. Mansfield will begin operations, and that section of country, like other localities where he has taken hold, will echo with the busy hum of industry instead of remaining in solitude and savagery (*The Ledge*, 14.12.1899).

The claims included the Gold Bank, Silver Queen, Harris, Yukon, Bostock, Jennie E., Northern Star, Josie Fraction, Marguerite, Maple Leaf, Rosene and Phoenician Fraction (*Wanganui Chronicle*, 23.1.1900). On the heels of this deal, Ernest brokered another large transaction which saw Rene Laudi acquire 38 000 shares of Excelsior Gold Mines and control of the company. Of the trio that originally staked the Joker and the Derby, Robert McGregor received $18 500 for 10 500 shares while A. G. Lambert and G. W. Taylor collected $52 000 for their 27 000 shares. Ernest was also said to be a large shareholder (*The Tribune*, 29.12.1899). At Christmas, Laudi cabled Mansfield instructing him to present all workers with a gift box. "His presents to his workmen will never be missed," said the Nelson *Tribune*, "but miners appreciate a man who remembers his workmen while 5000 miles away" (*Victoria Daily Colonist*, 28.12.1899).

Ernest planned to return to New Zealand in early 1900, where he expected to spend a few months and "acquire four miles of dredging ground for his London and French friends" (*The Tribune*, 29.12.1899). However, he changed his plans, instead examining the Molly Hughes mine near New Denver before departing for Europe (*The Ledge*, 11.1.1900 and 15.3.1900). In his absence, W. E. Boie was left in charge of Camp Mansfield, to "superintend the work, pay off the men, and generally have these important mining operations under his exclusive control" (*Spokane Daily Chronicle*, 19.1.1900).

Ernest on the rocks. Source: Wadd Bros. of Nelson BC, courtesy of Marion Everiss

One newspaper said Ernest's success in securing capital entitled him "to the distinction of being one of the largest mining operators in the province ... There is room in British Columbia for several score of Mansfields, and it is hoped that the practical manifestation of his confidence in our mineral wealth will not be without its influence in attracting others of them" (*Nelson Daily Miner*, 30.12.1899).

Ownership of the claims was becoming increasingly complex, as in 1900 the Joker and Derby were Crown-granted to Excelsior Gold Mines; the Treadwell, Philomene, Tony, Green Lakes, Marguerite and Alice fractional to Rene Laudi; and the Apex, Crescent, Twin Lakes and Green Lakes to W. E. Boie (http://minfile.gov.bc.ca/Summary.aspx?minfilno=082 FNW115). Ernest retained one-quarter or one-fifth ownership in most of these, either personally or through Excelsior. Workers were now stripping the Joker's ledge and tunnelling other claims. Good showings were reported from the Crescent and Twin Lakes, where 7- and 14-foot veins were tapped. Boie sent monthly reports to London (B.C. Minister of Mines Report, 1901). There were 50 tons of supplies at the mine, but Boie took up another four head of cattle, "which will suffice to keep the crew of 23 men going until summer" (*Spokane Daily Chronicle*, 19.1.1900). J. J. Fleutot was reportedly in Chicago and San Francisco arranging to buy a 20-stamp mill and cyanide plant to be installed in the spring (*Spokane Daily Chronicle*, 23.2.1900). "The Joker is practically a mine now and a prospect no longer. Its richness is undoubted and it will soon make the South Fork famous," *The Kootenaian* (1.3.1900) boasted.

Another delegation was appointed in Kaslo to head to Victoria and ask for $1000 to extend the wagon road to Camp Mansfield – touching off editorials in local newspapers arguing over where the road should begin (*The Kootenaian*, 22.2.1900).

A series of photos of Camp Mansfield taken by Wadds Bros. of Nelson around this time show the arrival of a pack train, the exterior of the Bertha tunnel in winter, the interior of the Bertha blacksmith shop and the interior of the Bertha and Marguerite cookhouses. Ernest appears in most or all of these. Also photographed were a general view from the camp, and Marguerite Falls, an attractive cascade nearby named after one of the claims. (Several of these photographs are held by the Touchstones Nelson Shawn Lamb Archives. In a conversation of 25 March 2010, Lamb said they were not sure whether Camp Mansfield was a mining or a lumber camp. Ernest was well and truly forgotten.) Welford Beaton described it this way: "Away up in the clouds lies Camp Mansfield, with the Sawtooth range of mountains shutting off the view to the east and the mighty Kitchener Glacier forming an everlasting icy barrier on the southwest. It is here that a little mountain brook trickles down from the miles of ice to gather strength as it tumbles over the rocks on its journey eastward, and soon becomes the raging and foaming south fork of

Kaslo Creek" (*Wanganui Chronicle*, 22.12.1899). That mountain brook had now become known as Mansfield Creek.

When Ernest returned in the spring of 1900, he announced $250 000 more to put into development work. In addition to Rene Laudi, his principal backers were French capitalists C. Heidsneck and E. E. Changeux of Reims (*Nelson Daily Miner*, 20.10.1900).

Ernest diversified his holdings by paying $25 000 for the assets of the West Kootenay Brick and Lime Co. Ltd., including brickyards at Balfour and Nelson and a quarry opposite Kaslo, where he planned to install new equipment. Ernest renamed the company the Mansfield Manufacturing Co. and assumed active management a few months later. He sent a "beautiful piece of marble and building stone" to an exhibition in Spokane, the closest US metropolis, and supplied pressed brick and terracotta for the KWC block, Nelson's largest and most prominent commercial building (*The Kootenaian*, 31.5.1900 and 13.9.1900; *Nelson Daily Miner*, 18.9.1900; *The Ledge*, 7.6.1900; *Henderson's British Columbia Directory* 1901; business card file at Touchstones Nelson Shawn Lamb Archives; and Pat Rogers, http://whenthewallstalk.blogspot.com/2009/09/kirkpatrick-wilson-clements-block-baker.html). West Kootenay Brick and Lime's president, W.W. Beer,

To left: Mansfield Manufacturing advertisement from 1901 & 1902. Source: Nelson Tribune. To right: Marguerite Falls and Ernest. Source: Wadd Bros. of Nelson BC, courtesy of Marion Everiss

was also vice-president of Pactolus Gold Mines Ltd., which was working claims in Camp Mansfield. A note in the *Journal of the Northwest Mining Association* (1900), discovered on Google Books, indicates that Mansfield Manufacturing won a silver medal in an ornamental stone competition.

Ernest also shipped 100 tons of ore from the Kaslo quarry to the Hall Mines smelter in Nelson as a test. The smelter was seeking a steady supply of limestone for fluxing, while Ernest wanted to know whether it had any gold value (*The Kootenaian*, 1.11.1900 and 27.12.1900).

"If the order is placed it will mean that a much larger force of men than is at present engaged will be put to work and that operations will be rushed throughout the winter which has not hitherto been done during that season of the year," wrote *The Kootenaian*. "In addition two new kilns will be constructed at the quarry and a large force of men will be engaged in getting out wood" (*The Kootenaian*, 1.11.1900).

Ernest stayed in Nelson at the opulent Hume Hotel and kept his office near the corner of Baker and Kootenay streets in a block he bought from Thomas G. Procter for $6000 (Henderson's British Columbia Directory 1901, Notes in business card file at Touchstones Nelson Shawn Lamb archives, and *The Kootenaian*, 18.10.1900). Here he opened a "mineral bureau" to exhibit specimens, including a piece of ore from the Joker that assayed three oz. in gold per ton (*Nelson Daily Miner*, 26.9.1900).

Expedition to Kitchener Glacier

After spending the early part of the summer in Europe, Ernest agreed to lead a sightseeing expedition on what he had called Kitchener Glacier. He and three others, including Welford Beaton, outfitted at Slocan City. Beaton said Ernest "rode a beast that looked like a superannuated English hunter" (Beaton 1.1902). They followed a pack train of 20 horses bound for Camp Mansfield, laden with tools, blasting powder, blankets and other supplies. After crossing the first range, they stopped for lunch and then pressed on to higher ground in increasingly heavier rain and snow, which stuck to their clothes.

They zigzagged up a narrow trail and finally rounded a bluff that brought them among the mountain peaks. It was August, but snow lay everywhere. The wind stung their faces and froze their clothes. "But notwithstanding the cold, the darkness, and our wretchedly uncomfortable condition," Beaton wrote in an account for *Canadian Magazine*, "we turned in our saddles and gazed on that magnificent scene." Half an hour later they were sitting before a fire in a miner's cabin, whose owner provided them with dry clothes.

Ernest in white jumper. Source: Wadd Bros. of Nelson BC, courtesy of Marion Everiss

The weather was better the next morning as they climbed even higher. Over the third and last summit, they began a precipitous descent on a narrow, rocky path that clung to the side of the mountain. By noon, they reached Camp Mansfield, tired and hungry, "feelings in which, I have every reason to believe, the horses were in entire sympathy with us. Their work was over, for the glacier could only be attempted by man." It stormed again that afternoon and all of the next day, affording them only glimpses of the glacier. During a break in the weather, Beaton tried to reach the lower edge of the immense ice field. After much slipping and sprawling, he succeeded, discovering a cavern along the way.

The next day offered perfect conditions and the opportunity for their ascent. Armed with snowshoes, snow-glasses, storm-caps and a little spaniel named Patsy, they headed for the eastern side of the glacier. "In two hours we … reached the foot of a wall of ice and snow. Here we poked steps in the hard crust with our snowshoes and Mansfield crept up," Beaton wrote. "We had tied the strings of our snowshoes together, and the line thus made reached to the first ledge." Patsy was hauled up by his collar. In another hour they advanced 30 feet. The extreme exertion kept them warm despite the bitter cold. As they gazed down crevasses hundreds of feet deep, they saw nothing but walls of intense green ice. Long detours were required to avoid such obstacles, but they made good time until they reached a huge drift on the north side of the peak, forcing them to move around to the steep south. "We sheltered ourselves from the biting wind behind a hummock of ice, lit our pipes, and had a good rest before we made the dash," Beaton said.

They were now 10 000 feet above sea level and breathing was getting difficult. They had to stop every few steps as they scrambled from ledge to ledge. A slip, Beaton noted, would mean instant death. During a rest, they wondered what had possessed them to do this. Lacking any satisfactory answer, Ernest declared if he got down alive, he would devote himself to climbing prairies instead. Two hours later, they were finally at the top. Beaton recalled:

> Then were we rewarded for the dangers we encountered and the exertion we exercised. No pen in the world and no brush could do justice to the scene that met our eyes ... There was nothing but scenery, grand, glorious scenery all about us. Beginning at our feet, it was without end — one vast sea of wondrous grandeur, with the motionless white-capped waves sparkling in the mid-day sun ... Our eyes sweep the horizon for 300 miles or more in every direction, but we see nothing but the untarnished whiteness of snow-capped summits that stretch upward their jewel-bedecked brows to be made glorious by the rays of the dazzling sun; and in all that region there is not a stir, not a sound. It is awe-inspiring in the intensity of its stillness.

On the return trip, they made sleds of their snowshoes and covered a thousand feet in an instant as Patsy yelped after them. They reached camp at sunset.

Behind bars
"Ernest Mansfield in jail" screamed the headline in the *Nelson Daily Miner* (19.10.1900*)*. "The English capitalist and promoter arrested and confined to a cell." "Arrested at instance of miners" said the Nelson *Tribune* (20.10.1900). "Ernest Mansfield, the well-known mine promoter, occupies a cell since Wednesday night."

Even more remarkably, Ernest had arranged for his own imprisonment.

This state of affairs followed a difficult summer at the Joker. The property suddenly shut down and was allowed to fill with water, as development was held up

> **Nelson Daily Miner, October 19, 1900**
>
> # ERNEST MANSFIELD IN JAIL
>
> The English Capitalist and Promoter Arrested and Confined in a Cell in the Provincial Jail--A Matter of Wages Due Men Working at Camp Mansfield.

Source: Nelson Daily Miner

for several months, much to Ernest's frustration (*The Tribune*, 20.10.1900). To that point, the work consisted of a 75 foot vertical shaft and 600 feet of horizontal tunnels (*The Kootenaian*, 20.6.1901). Ernest was further incensed having been given a favourable report from J. J. Fleutot, which indicated the Joker was "a regular bonanza." He suspected a complicated conspiracy to defraud smaller shareholders for the benefit of investors who held preferential shares. As a result, "all work was immediately stopped, ore dumped down the shaft, thrown in the creek, pumps and pipe hauled up, and the property injured in other ways" (*The Tribune*, 20.10.1900). Furthermore, although he thought he had arranged for enough cash to keep things going all winter, his French investors were evidently getting skittish. They cancelled a trip to Kaslo to inspect the properties, and further, "in view of the Boer war and the trouble in China they found some difficulty in taking money from their other investments to place it out in British Columbia" (*Nelson Daily Miner*, 20.10.1900).

By now two more companies were at play: United Gold Fields of British Columbia Ltd., created in September 1900 as the successor to Excelsior Gold Mines, and Mansfield Gold Mines Ltd., capitalised at $1 million, including $200 000 in working capital. Ernest's interest was about $125 000 and it absorbed one-fifth of his properties (*Nelson Daily Miner*, 24.10.1900).

As a large shareholder in United Gold, he planned to confront the directors for the right to re-open and empty the Joker and was willing to spend $30 000 of his own money to do so. "I will pay it myself," he said, "and as my object is to prove the property and ascertain its real worth beyond a shadow of a doubt, there can be no good reason why the directors will refuse …" (*Nelson Daily Miner*, 19.10.1900). He booked passage for England to sort things out, promising to resume work within two months. "Then you will hear of a property that will pay dividends, not for a year or two with a small plant, but for many years, and it will be a big concern..." (*Nelson Daily Miner*, 20.10.1900).

In the meantime, work continued on other Camp Mansfield properties, but the men were not paid. As the days turned to weeks, they became disgruntled and finally walked off the job. The miners were collectively owed $2000 while accounts for supplies also remained outstanding. Ernest discussed the situation with the workers, among whom he had always been popular. He cabled Rene Laudi, urgently requesting the wages, and asked the men to leave the matter with him a few days. When the money failed to show up, he hatched a scheme to make the company take notice.

One afternoon, he stood outside the Nelson provincial jail as a reporter walked by.

"This is the jail here, isn't it?" Ernest asked.

"That's the place where they put the bad ones," the reporter replied.

"Looks comfortable enough," Ernest said. "By the way, come up to my office about 5 o'clock and I'll have a good story for you."

The reporter did as he was told but found Ernest absent; the sheriff had taken him away. The unpaid workers had issued a *capias* for their wages – based on Ernest's advice.

"The entire incident has caused a mild sensation in Nelson," *The Tribune* wrote, "where Mansfield is well known and respected. It will be satisfactory to Mr. Mansfield's friends, however, to know that he is not repining under the circumstances, and that the situation, when thoroughly explained, will not redound in the slightest to his discredit." Normally a lien could have been placed on Camp Mansfield, but Ernest was gambling that putting himself behind bars would speed up the payroll. It was a means to an end, he reasoned. A day or two in jail would be worth it if the men were paid. Several prominent citizens offered to post bail, but he refused.

"It is hardly fair on you," a reporter visiting his cell said.

"There's lots of things that are not fair," Ernest replied, "but sometimes it pays to take strong measures to achieve tangible results ... Had I stayed out, the delay would have only continued and [I] hate delays of any kind, and more especially when it involves workmen's wages."

Asked about his accommodation, Ernest said it was "A-1. When I got located in my new diggings, numerous friends came to see me, and urged me not to remain, but I was firm and decided otherwise. I considered that punishment was due, and as the party deserving it was not here, I, in the interest of my workmen, resolved to take the medicine."

"Do you like the new bill of fare?"

"It's not bad, for a change at any rate, but I should not care for it very long. I must say, the porridge is a dream of liveliness ... and the bread perfection one would travel miles to get. But I have ceased taking the usual board, for after a day's fair trial, taking every meal as provided to the guests of Her Majesty, I must say, that I prefer the Hotel Hume fare" (*Nelson Daily Miner*, 19.10.1900 and *The Tribune* 20.10.1900).

Over the next several days, Ernest continued to hold court, receiving guests and smoking expensive tobacco while waiting for the syndicate's money to come through. The Hume Hotel sent meals over, and his barber paid a professional visit (*Nelson Daily Miner*, 20.10.1900). Ernest initially expected to be released within 24 hours, but as there remained no sign of the money, he could be forgiven for starting to worry. Outwardly, at least, he showed no break in his resolve. He stayed cheerful, and as the weather was miserable, consoled himself with the knowledge that while others were "compelled to walk about in the rain and mud", he was "dry, warm, and fairly comfortable" (*Nelson Daily Miner*, 19.10.1900).

The self-imposed jail sentence was a major inconvenience, however, for it prevented him from sailing for England as scheduled. But even with this he saw a silver lining: "Although I lose the *Majestic*, sailing next Wednesday, I shall certainly catch the *Oceanic*, sailing the 31st inst. It's a better boat you know, and I like comfort when I travel" (*Nelson Daily Miner*, 19.10.1900). Arrested on a Wednesday, the money had still not arrived by Saturday, meaning he would have to stay the weekend. On Sunday, a windstorm blew down all the telegraph wires into Nelson, and they were still out on Monday. Yet Ernest enjoyed another good night's sleep and a sumptuous breakfast and made the other prisoners envious with the smoke from his pipe *(Nelson Daily Miner*, 22.10.1900). He had another consolation: he considered the syndicate liable for his incarceration and intended to sue for damages (*The Tribune*, 23.10.1900).

Finally, a cable arrived from Lille, France stating that the outstanding wages had been wired to the local government agent. But the message did not arrive until after banking hours, so Ernest spent another night in jail. The cable also contained a peculiar offer: the syndicate wanted to buy a large percentage of his Joker shares at twice their price on the London stock exchange. Ernest refused (*The Tribune*, 24.10.1900). The next afternoon, the jailer informed Ernest that after almost a week in cell 33, he was finally free.

"Well, I wish you would wait until I finish my lunch", Ernest replied. "What if I refuse to go?"

"We will have to eject you for non-payment of board," said the jailer.

Ernest walked across the street to the Hume Hotel, where he registered as "Ernest Mansfield, Her Majesty's Prison." He was greeted with congratulations and handshakes from numerous friends who agreed it was "a damned shame" that he had been imprisoned and that he "ought to go back and raise Cain". He proceeded next to his office, where "several friends were tasting of the choice wines and cigars he keeps." To a reporter he praised the courtesy of his captors, pledged to re-open the Joker and announced he would sever ties with the syndicate while suing them for £1000 in damages (*Nelson Daily Miner*, 24.10.1900). In addition his suit asked for $65 000, of which $14 000 he said was due him for advances to cover the work on the claims, and $5000 for breach of contract for not forwarding funds which resulted in his imprisonment (*Victoria Daily Colonist*, 25.11.1900).

Although the bulk of United Gold Fields' money was French, it was registered as an English company, and Ernest intended to bring an action under the English Companies Act to compel them to reopen the Joker, claiming the value of his stock had been materially injured (*Nelson Daily Miner*, 20.10.1900 and *The Tribune*, 24.10.1900). Several people in Nelson offered to advance him $1000 on damages, possibly as an investment on any award by the courts, but he declined (*Nelson Daily Miner*, 24.10.1900). The outcome of Ernest's legal battle is not known.

Without naming names, he blamed his difficulties on "the jealousy of small minded people … Nothing seems too mean for these parties to say. Their object, I suppose, is to divert the capital I represent to themselves, but in this they will not succeed … The funny part of it is this: the people who go out of their way to attend to my business are indebted to me for the present position they hold. Biting the hand that fed them" (*The Kootenaian*, 18.10.1900).

Charles Plowman

At the Joker, a sack of ore was found in the lake near the mine. Brought to Kaslo, it assayed as high grade. Ernest declared he would take the sack to London to show the company directors as proof of the chicanery he believed was afoot (*The Kootenaian*, 1.11.1900). With Charles Plowman's help, Ernest obtained another £50 000 in backing within three weeks.

Plowman was one of Ernest's given names, so it is possible Charles was a cousin on his mother's side. Not much is known about Plowman, however. He appears to have been in the region by December 1898 as according to *The Kootenaian*, (8.12.1898), an H. Crawford signed an agreement with a C. Plowman, but he did not appear on the voters list that year. Nor was he on the Canadian census for 1901. Plowman was also involved with the Pactolus Gold Mining Co. and along with W. W. Beaton, met with the local Member of Parliament to convince him to finish the South Fork wagon road to Camp Mansfield (*The Kootenaian*, 24.5.1900 and 7.6.1900).

Plowman bonded a group of gold claims on Lyle creek near Whitewater called the Fletcher and left for London with samples. Assays showed the rock ranged

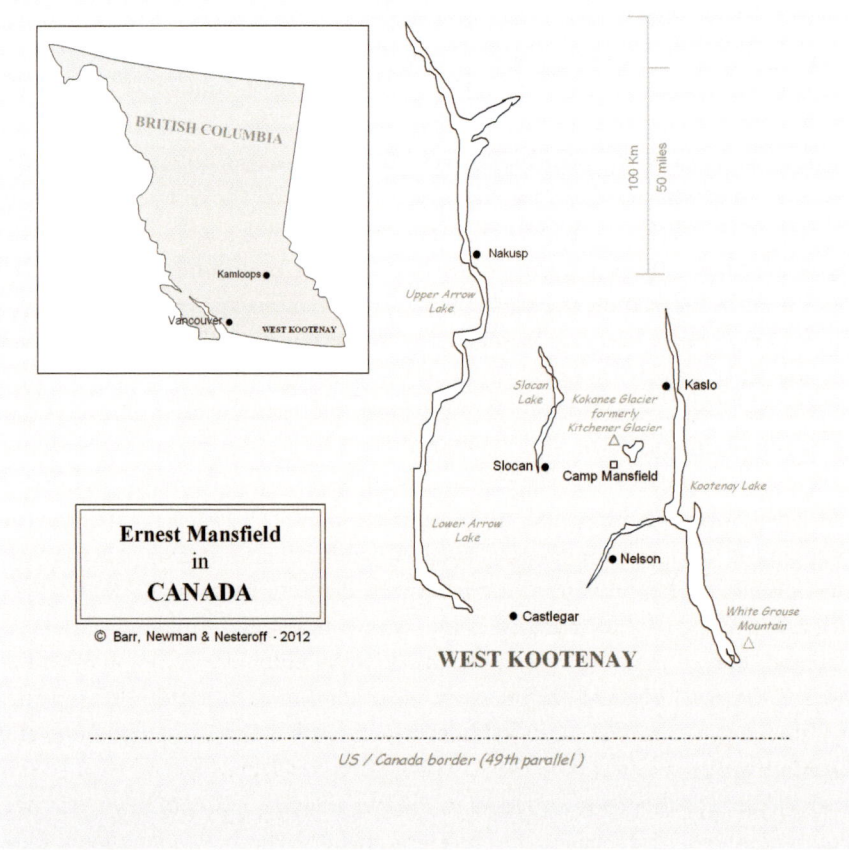

Produced by David Newman

from $29 to $230 in gold to the ton (*The Kootenaian*, 23.8.1900 and 29.11.1900). Within three weeks, Plowman secured £50 000 in European capital for Camp Mansfield and the Fletcher group. Ernest expressed "his unbounded admiration for Mr. Plowman's rustling qualities" (*The Kootenaian*, 30.8.1900). Although Plowman made several payments on the Fletcher, he evidently had some difficulty. A reporter asked him "You are finding it hard work to put this deal through are you not?" He cryptically replied: "Well yes, I must admit that, but not nearly so hard as it must be for those who are doing all they can to prevent it" (*The Kootenaian*, 14.3.1901). The Fletcher does not appear to have shipped any ore.

Leaving B.C.

In late 1900, Ernest formed a new company, Mansfield Gold Mines of British Columbia Ltd. (*The Kootenaian*, 6.12.1900). More bullish reports appeared insisting Camp Mansfield would be "opened up on a larger scale than ever" and that the Joker would be "worked in a manner that will prove its richness" (*The Kootenaian*, 7.2.1901).

Ernest returned to British Columbia in June 1901, intending to renew development at Camp Mansfield. He hired a Chicago firm to install a pumping plant and drain the Joker (*Nelson Daily Miner*, 9.6.1901 and 18.6.1901). Accompanying him was a new financier, Count Frederic de Baillencourt, who with his brother owned some of the largest linen mills in France. He planned to spend three months in the Kootenays, hunting, fishing and looking after mining interests (*Nelson Daily Miner*, 18.6.1901). Flush with cash – or hoping to appear so – Mansfield and Baillencourt donated $50 and $25 respectively to Dominion Day celebrations in Nelson (*Nelson Daily Miner*, 21.6.1901).

Whatever happened at the Joker that summer was not what Ernest hoped. Work did not proceed as expected, and in August he quietly left for England – never to return, although he may not have realised it (*Slocan Drill*, 9.8.1901). Shortly before departing, he announced plans to erect a marble building in Nelson, but this did not come to fruition (*The Ledge*, 4.7.1901).

The following year, he formed yet another company, the Kaslo-Slocan Mining and Financial Co. Ltd., capitalised at £500 000 to assume the idle properties at Camp Mansfield. This included both those that belonged to Rene Laudi and the Mansfield syndicate, plus the adjoining Chapleau and Kilo mines on Lemon Creek, whose previous owners installed a stamp mill and tramway. The new company promised to restart the Chapleau and the Joker (*Victoria Daily Colonist*, 3.11.1901 and 18.5.1902; *The Ledge*, 22.5.1902). But while annual assessment work was completed, the Joker remained closed *(Slocan Drill*, 29.8.1902). Meanwhile, an "exceptionally promising strike" was made on two other of Ernest's claims, the Black Hawk and the Daisy, and samples were sent to Europe *(Slocan Drill*, 3.7.1903). Then all went silent for several years.

After Ernest's departure

In 1907, the Kaslo-Slocan Mining and Financial Co. issued a new share offering although, having by now turned his attention to other pursuits, Ernest's involvement was probably peripheral at best as a stock certificate in the author's collection is signed by Count Baillencourt as a director, and by the assistant secretary, whose signature is illegible, but is not Ernest's *(*Kaslo-Slocan Mining and Financial Co. stock certificate 11.4.1907).

Two years later, three French capitalists, Count de Chantremerle de Villette, Count de Ferrand de Laurizen and Count de Ballincourt-Comol (presumably the same man as Baillencourt) visited the still-dormant Joker and liked it so much that they secured the property. Through the Selkirk Mining and Milling Co. they already operated the Cork and Florida mines up the south fork of the Kaslo River. The following year they were reported getting the Joker in shape (*Nelson Daily News*, 30.7.2009 and *Victoria Daily Colonist*, 15.4.1910). A new company, the West Kootenay Mining Co. Ltd., with offices in London and Paris, was formed to assume the assets of the Kaslo-Slocan Mining and Financial Co. Some surface and installation work was done in July 1910, with plans for a Pelton wheel and compressor to be installed the following spring. But for all of this, the company had no more success than any of its predecessors. It no longer existed as of 1921 (Mines Report, 1910, Stock certificate 30.8.1911).

As for the other companies involved in the Joker, Excelsior Gold Mines was voluntarily wound up and liquidated in 1901. Klondike Champs d'Ore Syndicate Ltd. was liquidated in 1911 (*London Gazette*, 8.1.1901, 23.4.1901, 13.9.1901 and 2.6.1911). In the latter year, Ernest reportedly had "an extremely favourable offer from a powerful European financial group to go out again to British Columbia" but declined because he was by then too busy with prospects in Svalbard (NEC Prospectus 1911/12). In the end, Ernest had little to show for his British Columbia excursion. What money his successful operations made, he ploughed back into less successful ones. As he noted years later, "From Klondyke I went to British Columbia, where I had a tidy stay. I made a couple of fortunes and got through them. Don't ask me how: for it's the same old story – and I suppose will always be with the likes of us." (Mansfield 1910).

Little remains in British Columbia to remember Ernest Mansfield by, and indeed he is no longer well known, even to historians. It is as if his memory has been deliberately erased. For some reason, none of the place names he was associated with stuck: Kitchener Glacier was renamed Kokanee Glacier, returning to the name it had prior to Ernest's arrival. (http://env.gov.bc.ca/bcgn-bin/bcg10?name=6220), and in 1922 it was declared a provincial park, including all of Camp Mansfield. One of the peaks is known unofficially as Mount Kitchener, but its official name is The Pyramids (http://archive.ilmb.gov.bc.ca/bcgn-bin/bcg10?name= 22197). Camp Mansfield is no longer on any map, although the proposed mill site is now the parking lot for a popular hiking trail to the Joker Lakes (which, although next to each other, are different colours – one a deep blue, the other a milky turquoise).

Mansfield Creek is now Keen Creek. Even Marguerite Falls, whose name Ernest may or may not have coined, has been rechristened Bridal Veil Cascade. No ore was ever shipped from the Joker mine, although the reason remains a mystery. Field trips by geologists in the 1920s and '30s noted two old cabins on the John A claim used by visitors to the park (one still stood, derelict, in the early 1980s, a short distance from the upper lake [Mines Report, 1933; Cairnes, 1935; Carter and Doug Leighton, 1980]). Many old tunnels were also noted, some of which were starting to cave in. A report observed one "was not driven far enough" while another "did not reach its objective." Assays returned gold values of up to 2.7 oz. per ton (Mines Report, 1933).

"To sum it all up, the Joker was never a mine — just a prospect," historian Don Blake wrote. "They had big plans for the construction of a mill at the Joker mill site, but these plans fizzled out. It probably was more of a stock promotion than anything else as they had very little ore reserves to build a mill for." (Blake 1988). Nor, it seems, did anything become of the White Grouse claims Ernest bought from Jennie E. Harris at considerable cost. While he received an extension on the property and indicated plans to "organize a strong company and commence extensive operations", there is no sign that it happened (*The Ledge*, 20.9.1900). Yet in a positive reflection, perhaps, on Ernest's character, he was never blamed for these failures.

Today on the site of his brickyard in Nelson is an apartment building. The KWC block, for which he supplied materials, still stands at the corner of Ward and Baker streets. Another surviving building, originally the post office, then city hall, and now a museum and art gallery, was built in 1902 using material from Ernest's marble quarry, although it is not known if he still owned it at the time. The Hume Hotel, where he kept his quarters, is still in business, although unsympathetic renovations in the 1920s robbed it of much of its grandeur. The jail where he spent a week doing penance in hopes of securing his workers' wages has long since been demolished. Mining in the West Kootenay has been eclipsed by forestry as the major industry, and even this seems perpetually on life support. Nelson itself has no apparent means for continued existence but remains a major trading centre, thriving on tourism and an underground marijuana trade. Its downtown remains in an excellent state of preservation, thanks to a heritage conservation programme of the 1980s.

Ernest's partner-turned-nemesis J. J. Fleutot and United Gold Mines are actually a little better remembered. Despite the company's name, in 1901, Fleutot and a colleague went looking for coal in the Crowsnest Pass of Alberta. They discovered a seam and founded the town of Lille. The company was renamed Western Canadian Collieries in 1903, and Fleutot served as managing director. He ultimately returned to France and died in Toulouse in 1919 (*Historical Archaeology*). Lille is now a ghost town but has been declared an historical landmark (http://coalking.

ca/people/ghost_lille.html). Western Canadian Collieries still owned some of the claims in Camp Mansfield as of the mid-1930s (Cairnes, 1935).

One of Ernest's relatives followed in his footsteps in British Columbia, although he did not know it at the time. Grandnephew Ross McNeil began visiting the West Kootenay in 1979 and lived for two years in the 1990s at Balfour – where Ernest once owned a brickyard. McNeil visited Kokanee Glacier Park several times, unaware Ernest trod that ground nearly a century before (McNeil, 29.12.2010).

The NEC logo from their two Svalbard prospectuses. Courtesy of Richard Gardner

ERNEST IN ENGLAND
(1904–1924)

THERE IS A GAP from mid-1901 to Mansfield's arrival in England in 1904 and it has not been possible to fill it with any certainty. At this point, therefore, this remains one of the mysteries of his remarkable life. However, after travelling all around the world it is certain that Ernest Mansfield returned to the UK in 1904 at the age of 42 to reside in the village of Goldhanger in Essex. The local medical general practitioner Doctor Salter informs us in his published diary (Thompson 1933) that Mansfield and "his nice fresh pretty-looking Scotch wife" took a cottage at Goldhanger and that the doctor himself delivered their baby Bernice Zoe there. There have never been any gold mines in or near Goldhanger despite its name, so the reason for coming to Essex can only be surmised: It is known that Mansfield went to London in 1899 to attend "The Greater Britain Exhibition" with a mining colleague, a Mr Timmins. We know that the sister of Margaret Mansfield, Bella Booth, later married Mr Timmins and it is also known that the Goldhanger rector, the Rev. Frederick Gardner, had professional connections with a Rev. Timmins and that one of his daughters, Angela, later married a Rev. Timmins. So it seems very likely that it was a combination of family and professional connections that brought Ernest to Goldhanger.

Baptismal records in St Peter's Church, Goldhanger, reveal that the baptism of Bernice Zoe Mansfield took place in June 1904 and the ceremony was conducted by the Rev. Gardner, who later went to Svalbard several times with Mansfield. The birth of Bernice Zoe was registered at the Maldon Register Office in Essex UK, using the English spelling of Zoe (birth certificate (20.6.1904). The mother's maiden name was given as Margaret Booth with the address of The Limes, Goldhanger. The house called "The Limes" at that time was owned by the Rev. Gardner and

The Old Parsonage in Goldhanger in the 1930s. Source: Goldhanger Digital Archives

The Old Parsonage in Goldhanger in 2010. Photo: David Newman

was subsequently renamed "The Old Parsonage" and was used as the residence for the curates who were employees of the Rector. The Rev. Gardner's first journey to Svalbard took place in 1904. The Rector's second journey to Svalbard, this time with Ernest, was in 1905 (see the later section about the Rev. Gardner).

Dr Salter lived two miles from Goldhanger, in the village of Tolleshunt D'Arcy, and had been the general practitioner for a large rural area that included the village of Goldhanger. The doctor was extremely wealthy and travelled extensively in Europe, which he recorded comprehensively in his diaries (Thompson 1933). The doctor and the rector obviously knew each other extremely well even before Ernest arrived, and the partnership formed between the three men was pivotal in their exploration of Spitsbergen and the initial formation of their own company in 1905. It seems Dr Salter himself never went to Svalbard, although he undoubtedly provided significant financial and moral support (see the later section about Dr Salter).

Ernest with Charles Mann hunting off Spitsbergen. Source: Charles Mann collection, courtesy of Rosemary Mann

Charles Mann and George Alexander hunting off Spitsbergen. Source: Charles Mann collection, courtesy of Rosemary Mann

In the summer of 1906 the three partners employed two Goldhanger men to go to Svalbard with Ernest to build accommodation huts. They were Charles Mann, the Goldhanger village builder and undertaker who it is believed was paid £40 for the trip, and his friend George Alexander. These men lived just a few houses away from Mansfield in the village. Charles Mann and his family owned a wheelwrights and cycle shop in the village square. George Alexander's family business was described in a trade directory of that period (Kelly's 1899) as "Beer Retailers & Shopkeepers" in Fish Street, still known today by its Alehouse name as "The Bird in Hand".

Charles Mann took a box camera with him on the trip to Svalbard, and the photographs remain in the possession of his family, some of whom still reside in the village.

In later life Charles Mann took a very active role in Goldhanger village and became chairman of the Parish Council, bell tower captain, school manager and landlord of The Chequers Inn. Nine professionally produced photographs of Mansfield in a mining location are still in the possession of Charles Mann's family and these were previously thought by the family to be from Spitsbergen, but it is now known that these were taken at Camp Mansfield near Nelson, British Columbia (see the chapter *Ernest in Canada*). Mansfield is strategically situated in the middle of all of the photographs wearing a light coloured jumper. Charles Mann is not

Ernest in the middle. Source: Wadd Bros. of Nelson BC, courtesy of Marion Everiss

Ernest at the head of the table. Source: Wadd Bros. of Nelson BC, courtesy of Marion Everiss

Camp Bell cabin, Svalbard, built by Charles Mann and George Alexander. Source: Charles Mann collection, courtesy of Rosemary Mann

known to have ever visited British Columbia and he is not in the photos, so one can only speculate as to how they came to be in his possession.

Charles Mann made a second visit to Svalbard in 1908 for which he was paid £60, and some of his photos are annotated with the year 1908.

While based in Goldhanger, Ernest made many visits to Svalbard between the years of 1904 and 1911 and must have spent a considerable amount of time away from his home and family – although pictures apparently taken in Tromsø of him with his wife and daughter (Booth 1912) suggest that they travelled with him as far as Norway on at least one occasion.

The three partners registered the Northern Exploration Company (NEC) at Companies House in London in 1910 and a year later, in May 1911, Mansfield sold all of his rights to land on Svalbard to the Company for a fee of £75 000, to be paid in the form of shares. A sales agreement (Sales 6.5.1911) lodged with Companies House in London specified that Mansfield should "give his exclusive services to the Company for a term of five years".

In September 1911 a lengthy and impressive article, most probably written by Mansfield himself, appeared in various major newspapers around the world including: the UK *Daily Chronicle*, the *New York Times*, the *Wanganui Chronicle* (NZ), the *Grey River Argus* (NZ), the *Western Mail* (Perth Western Australia), *Barrier Miner* (New South Wales), the *San Francisco Call* and others. Ernest had left New Zealand 13 years before the article appeared in the *Grey River Argus* on 27 September 1911 (*Argus* 27.9.1911). This is included here in full:

Ernest, Margaret and Zoe (probably taken in Tromsø). Source: David Booth, courtesy of Ross McNeill

Zoe, Mr Timmins, Margaret and Bella (probably taken in Tromsø). Source: David Booth, courtesy of Ross McNeill

In Search of Gold

Tolleshunt d'Arcy, a quiet little Essex village, is the centre from which has been engineered a scheme having for its object nothing less than the exploitation of an Arctic goldfield and coal-field, the latter of which may supply Europe with fuel when its own coal has been exhausted.

There are three principal characters in this new "Treasure Island" story of real life. One is the Rev. Mr Gardner, rector of Goldhanger, a sleepy little village of Essex, whose greatest excitement hitherto have been a wedding, a birth, and a death. The second is Dr. Salter, of Tolleshunt d'Arcy, three miles from the rectory of Goldhanger. The third is Mr Ernest Mansfield, a musician, a man of letters, a great traveller, and a mining engineer, who is the neighbour and friend of the rector and the doctor.

After their day's work it was the habit of these three cronies to meet in one or other of their houses and to talk of their experiences over a glass of wine. Both Mr. Mansfield and Dr. Salter are men who have hunted and explored in wild places, and the conversation of Mr. Mansfield was especially interesting, because, as a mining engineer and one of the early pioneers of Klondyke, he had searched for gold in many parts. Always his conversation came back to gold, gold, gold, and the possibility of new discoveries. He held firmly to the theory that there were great gold deposits in the Arctic regions at present untouched by man. The Rev. Mr. Gardner was secretly fired by his words, and one day he said, "I am going to Spitzbergen. Perhaps while I am there I may put your theory to the test."

Sands of Gold

It was the clergyman who was the cause of the discovery which led these three friends in Essex to share an amazing secret. It seems strange enough that the rector of a rural parish should travel into the Arctic regions for a summer holiday, but stranger things than that were to follow. The Rev. Mr Gardner, acting upon the expert advice of his engineering friend, brought back from his voyage pieces of quartz and rock, and specimens of sand and mud from the Arctic coast. To him they were meaningless. He smiled as he thought of his strange baggage. But one night there was a thrilling sense of mystery and excitement when the three friends gathered round these little heaps of rubbish in the sitting room of the rectory. Mr. Mansfield pored over these pebbles and bits of rock, held them up to the light, and examined them closely.

"Well?" said his friends. "Gold," he said, "or I'm a Dutchman." Gold! It seemed incredible. Here in this little Essex parlour was a secret of amazing possibilities and importance. The specimens were sent to London to be tested. The report that came confirmed Mr. Mansfield's opinion. The sand brought back by the clergyman was what is known as "pay gravel," the washing down of a gold deposit. The three friends formed a private syndicate, and Mr. Mansfield went out to Spitzbergen to prospect more closely and take out a claim. He found that a party of Americans were in advance of him, but they entered into friendly relations, and the Americans went further up the desolate coast, where they have pegged out their own claim and have now established a small township engaged in coal digging with good results.

The details of what Mr. Mansfield found must still be kept secret, although it can no longer be hidden that there are the most astonishing indications of gold and an inexhaustible coal supply in this unexplored territory of the Frozen North. One great difficulty now faced the village pioneers. From whom were they to get the full right to take possession of minerals in this region? No flag of any nation flies over its barren rock. It is "No Man's Land." Dr. Salter approached the Foreign Office and obtained certain advice, upon which he is now acting. Upon Mr. Mansfield's return money was raised with the help of private friends in the neighbourhood, pledged to keep the secret. The doctor, the clergyman, and the engineer, have already made their way to the Arctic regions, and only the other week an iron-built vessel steamed out of the Thames bound for Spitzbergen, with a crew of English, Scottish, and Norwegian sailors and miners.

"We have now a very flourishing little mining colony on our Arctic claim," said Mr. Salter, "strong enough to defend themselves in case of need and ready to enforce that mining law which means death to anybody who tries to 'jump' a claim".

"The ships have taken out a great supply of provisions, and much is required to sustain the body of men utterly isolated from the world and depending for their lives upon what they have carried with them. They have built houses taken out in pieces and under strict discipline of mining engineers and officers they are leading a hard, lonely life, with plenty of toil, and no other society but their own. Around them they hear the barking voices of the seals who lie upon the rock ledges, polar bears prowl over the barren region, and no human being outside their own camp disturbs the utter solitude. They keep close to their huts, for the Norwegians especially are superstitious, and afraid of the ghosts which they believe haunt these desert regions." One day these pioneers in search of Arctic gold made a gruesome discovery. There on the naked rocks lay three skeletons. Their bones were bleached and stripped clean of flesh. There was no sign to show the race or character or history of the men who had perished in this Arctic solitude.

In 1911 or '12 the Northern Exploration Company published an impressive "Shares Prospectus" (NEC 1911/12) in London, which included testimonial letters from Ernest, Dr Salter, the Rev. Gardner, Charles Mann and George Alexander, as well as from many other company directors and experts. The prospectus was expensively produced as a hard-back book with "NEC" gilded on the cover and gold leaf edging and had 132 pages of text and photos. The original photographs are in Charles Mann's photo album (Mann 1908).

Mansfield's address in this Prospectus is given as "The Limes," Tolleshunt D'Arcy. This property is near to Dr Salter's impressive residence and was also owned at the time by the Rev. Gardner, so presumably it was no coincidence that both houses where Ernest stayed in Essex were called "The Limes."

While residing in Tolleshunt D'Arcy in 1913 and travelling back and forth to Svalbard, Mansfield wrote his second novel entitled *Ralph Raymond* (Mansfield 1913 – see the *Literature and Music* chapter).

In 1913 the Northern Exploration Company produced their second and equally impressive hardback shares prospectus book (NEC 1913) entitled: "Marble Is-

"The Limes" at Tolleshunt D'Arcy in the 1920s. Source: Goldhanger Digital Archives

land", which again included many testimonial letters from company directors and experts and many more photographs.

In 1915, during the First World War, the *Wanganui Chronicle* in New Zealand reported that "Mr Ernest Mansfield, an erstwhile Wanganui resident, was fighting for King and country". Unfortunately, no further information was given, and it has not been possible to find any details about this.

In 1919 the Goldhanger parish magazine (GPM 1919) reported that Mr & Mrs E. Mansfield and Zoe Mansfield had made contributions to the Goldhanger War memorial fund being organised by the Rev. Gardner, which is an indication that they still had close connections with the Rector and the village at that time.

Early in the 1920s Ernest Mansfield became Master of the Easterford Masonic Lodge, at Kelvedon in Essex, which is seven miles east of Goldhanger and Tolleshunt D'Arcy. Dr Salter had earlier been associated with the building of this Lodge. He had a long association with Essex Freemasonry and was at the time the Essex Grand Master. There is a framed illuminated scroll in the entrance hall of the Lodge which was presented to Dr Salter in the early 1920s to celebrate the Doctor's long association with the masons, and Ernest's name is among the list of contributors. The question of Mansfield's Masonic connections is considered in the last chapter of this book.

In 1923 Mansfield presented Dr Salter with a Marconi radio that at the time was the latest technology and cost £120. This would be the equivalent of £5 000 in today's terms and was an extremely generous gift.

The Reverend Gardner's involvement

The Reverend Frederick Gardner was Rector of Goldhanger and Little Totham from 1893 to 1936. The Rector was already 40 years of age at the time Mansfield came to the village and was not an entirely fit man. However, despite apparently suffering from what was recorded as motor neuron disease the Rev. Gardner was a great traveller and was probably a wealthy man as many rectors were in that period.

As well as the family and professional connections between Mansfield and Gardner, there is another quite independent reason for why the Rector became involved in the exploration of Spitsbergen. In a book on the history of the adjacent village of Little Totham (Key 2005), Lorna Key explains that the Rector stayed on the Earl of Morton's estate in Scotland in 1899 and that he may well have known the Earl before that time. It is known that the Earl had coal mining interests in Spitsbergen at around that time and it is also known that the Rector joined the Earl on sailing trips in the North Sea and to Svalbard.

Goldhanger and Little Totham villages are just two miles apart and have shared the same Rector for hundreds of years. In his 43 years as rector, the Rev. Gardner clearly had a large and long-lasting impact on both villages. *Little Totham – The story of a small village* has a full chapter devoted to the Rev. Gardner and these extracts from the book are reproduced with the kind permission of the author.

Rev. Frederick Gardner (1864–1936). Source: Marble Island prospectus

The Rev. Frederick Gardner was the last of the Victorian rectors. He came from a middle class family and was sufficiently well off to be able to pursue his interest in foreign travel and to appoint a chaplain in his absence. He was a benevolent and conscientious priest and had a caring affection for Little Totham. He was generous to the church and its people and his influence was very important to the life of the parish. It is for this reason that a whole chapter is allocated to him...

The Rev Gardner loved travelling. He was very interested in investigating unexplored territories in Northern Scandinavia, and used to obtain permits from the government of the day to visit Spitzbergen. His main interest lay in exploring the mineral wealth of the area and he always hoped to find something original and exciting, possibly gold. It was on one of these trips that he was accompanied by Charley Mann from Goldhanger who was a blacksmith engineer. These trips were curtailed from 1911 onwards when the First World War began looming.

His regular journeying around Europe must have been a lifeline for him and his wife and he continually sent messages back to his parishioners through the Parish Magazines – to whom such journeys must have seemed remarkable.

In September 1899 he wrote of the first signs of his illness: "My dear people, I have been amongst you for nearly six years and for the first time within that period I am about to take a prolonged holiday, partly under doctor's orders after an acute attack of rheumatism. I am to go as the private chaplain to the Earl of Morton for eight weeks at the seat of Congalen Ardgour in N. W .Scotland." From there he wrote telling his parishioners about the grandeur and the magnificence of the scenery, the people who attended the services and the highland games, all of which must have seemed part of another world to the people of his parish.

In August 1900 he left for Garmisch in Bavaria where he visited Oberammergau among many other places, and ended with a cruise down the Rhine. Afterwards he returned to Scotland for another prolonged visit. On all these trips he always left the parish in the hands of priests, often from distant parts, and the curate of the time.

In August 1906 he travelled to Spitzbergen in the North of the Polar Sea, having "a rough and somewhat dangerous time." A lantern slide show was promised on his return. In August 1907 he returned to Spitzbergen with Mrs Gardner, but they never reached it due to the large lumps of floating polar ice which rendered it impossible for the steamer to pass. By October of this year his condition had worsened and he was ordered to see a consultant in London with a possibility of arresting the complaint. He was in hospital until December and returned to Goldhanger after a period of convalescence in Bournemouth. In April 1908 he returned early from Biarritz in the South of France as his eldest boy was ill. In July 1908 he visited the baths at Oeynhausen, Germany, for a prolonged cure.

In January 1910 he was to be found 1,000 miles along the River Nile south of Cairo surrounded by the Great Desert. Post took nine days to reach England, but his letters then described in detail the pyramids and the tomb of Pharaohs, ending: "I have just returned from taking tea with the Canon of Jerusalem." In November 1910 he visited the sulphur baths in an effort to ease the illness. "Meningitis, or inflammation of certain tracts of nerves, is the root of my trouble, contracted through frequent chills and exposure." Nevertheless he looked forward to returning to a more vigorous life.

The Rev. Gardner made at least four trips to Svalbard, the first with his wife, as passengers on the steamship RMS *Ophir* in 1904 (Reilly 2009). Together with Dr Salter, the Rector played a significant role in the creation and development of the Northern Exploration Company. The Rector was a director of the Company and may well have invested substantial funds in the enterprise. The 1911/12 Prospectus has in it a long letter from the Rector, and the 1913 version has a photograph of him with the title: "early pioneer".

The Goldhanger & Little Totham Parish magazines (GPM 1919) of the period refer to four summer trips made by the Rev. Gardner in 1904, 1905, 1906 and 1907. The first two just refer to trips "to the far north".

The 1913 prospectus reveals that the Rev. Gardner's brother, Dr F. G. Gardner, also participated on an NEC expedition in 1912 as the company doctor and wrote a report which is reproduced in the prospectus. Seven years later in 1920 Doctor Gardner wrote a letter to *The Times* (18.12.1920) extolling the virtues of the Spitsbergen climate and its potential for further exploration.

In recognition of the Rector's role in NEC's early development, a small motor and sail-driven lighter based at Spitsbergen was named *Cynthia* after the Rector's daughter, Ethel Cynthia Gardner (1906-62). A similar boat had been named *Zoe*, after Ernest's daughter.

The Rev. Gardner died in 1936 at the age of 72. His obituary in *The Times* (1936) included this reference to his involvement in the Spitsbergen expeditions and the Northern Exploration Company:

> In the early years of the present century he conducted a series of expeditions to the Arctic, in which he was associated with the late Dr. J. H. Salter of Tolleshunt D'Arcy who gives some account of the venture in his "Diary and Reminiscences." On one occasion Mr. Gardner crossed the Arctic Ocean with one Englishman and two Norwegians in a 52ft. sailing boat which was wrecked on reaching Spitsbergen and all hands were rescued by an American whaler. His work there led to the formation of the Northern Exploration Company for the purpose of exploiting the great resources of Spitsbergen, but early expectations were never realized. He suffered many privations there which largely contributed to his lameness in later life. During the War he joined the late Lord Morton's yacht in a mine sweeping expedition operating on the west coast of Scotland.

Dr Salter's involvement

Dr John Henry Salter lived two miles north of Goldhanger, in the village of Tolleshunt D'Arcy, Essex, and was the medical general practitioner (GP) for a large rural area that included Goldhanger. The doctor was wealthy, well connected, and had travelled extensively in Europe (ODNB). His life is comprehensively recorded in

Dr John Henry Salter as a Freemason. Source: Dr Salter's Diaries

his published diary (Thompson 1933), and he remains very well known in the area because of that.

Dr Salter of Tolleshunt D'Arcy in the county of Essex: his diary and reminiscences from the year 1849 to the year 1932 was compiled and published in 1933. The biography consists largely of extracts from Dr Salter's own very extensive manuscript diaries. After his death in 1932 they were acquired by Alderman John Ockelford Thompson, proprietor of the *Essex Chronicle*, who published a small selection prefaced with 'An appreciation' by the fifth Earl of Lonsdale. In addition to the diary extracts are fifty-nine short, amusing, amazing and informative reminiscences (ODNB). Sadly, most of the originals were destroyed in a fire at J. O. Thompson's house in Chelmsford after it was struck by a flying bomb in World War II. It is the voluminous diaries, a remarkable record of social history, which make Salter immortal. Obituaries of the doctor appeared in *The Times* and the *Essex Chronicle* in April 1932, and there is a biography of Dr Salter in the Oxford Dictionary of National Biographies (ODNB).

The partnership formed between Mansfield, Gardner and the doctor was pivotal in their exploration of Spitsbergen and the formation of the Northern Exploration Company. Although Dr Salter does not appear to have ever visited Svalbard, he clearly provided significant financial and moral support to the other two men.

Dr Salter (1841–1932) was already 63 years of age when Mansfield arrived in Essex. During his lifetime he built up a reputation that embraced medicine, the military, horticulture, wildfowling, horse and dog breeding, sportsmanship, prize-fighting, a magistrate's role and freemasonry. Although married, the doctor had no children. Dr Salter attained great prominence in UK freemasonry. He was a

Dr Salter's house in Tolleshunt D'Arcy in 1910. Source: Goldhanger Digital Archives

Dr Salter's house in Tolleshunt D'Arcy in 2010. Source: David Newman/Goldhanger Digital Archives

founder member of the Easterford Lodge in Kelvedon, Essex, and became Deputy Provincial Grand Master and Grand Deacon of England. He reached the 32nd degree in the craft, the 33rd degree being reserved for crowned heads alone.

The following extracts taken from Dr Salter's diary are those which specifically refer to Ernest and the NEC

21 May 1906 Made a director of the Spitzbergen Exploration Co.

18 July 1906 Dined at the Goldhanger Rectory and heard Rev. Gardner's story about Spitzbergen, from which he had just returned. His account is wonderful and there seems to be great results looming over all participants, myself included.

20 Aug 1908 A telegram that the Spitzbergen assay was 16 grains per ton and therefore "useless".

13 July 1911 Letter came capsizing my plans for going to Spitzbergen. I am out of it regarding polar bear and reindeer shooting. My 70th birthday tomorrow too!

31 July 1911 Bombarded by newspaper men about Spitzbergen.

31 Jan 1912 First meeting of Northern Exploration Co, Spitzbergen. There were some splendid exhibitions of marble of all colours. It seems to me that things will now go.

25 Mar 1918 Clinched a deal with the Northern Exploration Co, receiving £1000 for 2100 shares. I think I am wise. [Note: Equiv. to £200 000 today]

10 Nov 1918 Spitzbergen resuscitation (after 24 years)!

6 Feb 1920 Spitzbergen property seems to have done well after all. I am the poor chap left out in the cold!

10 Mar 1923 To Marconi House (Chelmsford) to see the installation that Mansfield has ordered to be put into my house at a cost of £120. [Note: Today £120 would be the value of a house]

Much of the chapter from Dr Salter's Diary entitled *Reminiscences* is included here, while the remainder is found in the chapter *Ernest in Svalbard*:

Gold in Spitzbergen

At the end of 1904 a man and his wife came and took a cottage at Goldhanger, a village near here. The lady was expecting to be confined with her first baby, and I was asked to attend her. She was a nice, fresh, pretty-looking Scotch girl, and her husband was a man of extremely notable appearance. He was a man you would pick out in a crowd as being somewhat remarkable – a strong, well set-up, muscular man perhaps a little more than middle height, and evidently well versed in the things of this world. I suppose at that time he might be round about thirty. He had a most interesting face, dark, crisp, curly hair, a nice, open, intelligent dark-grey eye, good teeth, a well-shaped nose – not Jewish, but going to a point – and a good, strong jaw.

His manner was pleasant. He was talkative, and at times inclined to talk somewhat at random so that you could not quite think that all he said was free from exaggeration. However, he said what he was – an adventurer, a mining prospector, had travelled a great deal under some very rough circumstances. Among other things he had taken a share in discovering the Wei Hei mines in New Zealand [Note: Probably Waihi mines]. He was also one of the few men who got safely to Klondyke. He went right across Australia in the dry season, a thing that had never been done previously. What was more, he was accompanied by another man whom he carried on his back during the last three days. I mention that because Mansfield – that was his name – was a man in every sense of the word. He always went for big game, but, as I say, to our sublunary minds he was given to exaggeration, and it might be that we were not educated up to it, for when he said anything extraordinary it seemed too extraordinary for us to understand.

His wife went through her confinement. It was a long and troublous time, but those were the days when the medical practitioners stuck to things themselves and would not hand them over to hospitals or homes. The child was a girl, and was worshipped by her parents, whom I came to reckon among my most devoted friends.

But I can't say that I took to Mansfield at first – he was so full of talk, and apparently of swagger, that I could not believe it was all real. For instance, in the morning after I had been there the whole night and had had enough to think of to turn a man's hair grey, Mansfield, who I suppose treated a confinement in the hands of a doctor as an everyday matter, was busy already about the house making it tidy, and as I went downstairs he greeted me with, "Now, Doctor, what is it to be, champagne, whisky, or what?" when all I wanted was a wash and to go to bed.

But after a bit I got to know his valuable points, and I believed most that he told me. His geese were always swans, but I gathered that that is characteristic of prospectors for gold and so on. Their view seems to be proportionate to what they are looking for – copper is gold, crystal is a big diamond, and so forth. Well, one day I asked Mansfield to come to dine with me, and he came, and a parson came with him. This parson was then under my care [Rev. Gardner, suffering from motor neuron disease]. I was recommending him to have a rest, and at this dinner-table we were talking of the places he might go to. Among others was a visit by one of Lunn's ships to the land of the Midnight Sun – up to Norway. As soon as I mentioned the Midnight Sun Mansfield got excited and said, "I wish you'd send me there, too, because all the gold that is undiscovered lies around the polar regions, and I'm sure there's a lot about the Arctic Circle, just as there is at Klondyke. That is where I'd go if I had the money."

The divine warmed up at this, and they talked about it. Mansfield said he would fit the parson up with a miner's outfit, which meant a bag or sack containing certain tools adaptable to a man to carry on his back certain distances in prospecting for minerals. He would also tell our mutual friend where to go to make the best use of his time.

"When you get up there in Spitzbergen," he said, "the snow will have melted, the sun will have bared the land, you will find a carpet of flowers everywhere, and the rivers, which have been roaring torrents during the late spring and early summer, will have discharged into the sea and will be fordable and comparatively empty. There is no end of these rivers, but you must remember this – that they have brought bits of rock long distances, which you will find at the corners. At the corners also you will find mud, which is disintegrated rock. That has been brought some distance, perhaps not so far. At any rate, make your way up some of these rivers, and into your sack put portions of this gritty stuff and mud and bits of rock, and when the other people are careering about eating cakes and dancing jigs under the Midnight Sun you will be using your time to much greater advantage. Bring your specimens home with memoranda as to where you got them, and we will see what they contain."

We waited with a good deal of interest, and it was all done. We sent specimens of the bits of rock and of the mud to the principal assayers of the Bank of England, and asked them to give a report upon them. They found gold in small quantities in them all!

Mansfield then said, "I'm off myself now." I said I would like to go also, but I had unfortunately to stay at home professionally, and, of course, you can only go to Spitzbergen when the seas are open. So I was left out of the picture. The other two went across by steamboat to Norway, and there hired a whaleboat in which they went up to our island. They went into the big Bell Sound there and they landed. They stayed prospecting, making maps, building cairns to show the localities, burying bits of parchment and so on to show each place that they had investigated. They found coal in large quantities, coming out of the rocks horizontally, so that you had nothing to do but pick your coal out, roll it down the mountainsides, and load it up in ships in deep water at the foot. I said, "Give me the coal, and you can have the gold."

The remainder of this passage concerning the NEC is quoted in the chapter *Ernest in Svalbard* under the headings *Ernest winters 1908–09 by Bellsund*, *The 1911 Expedition* and *The NEC is dissolved*.

Mansfield's first book, *Astria – The Ice Maiden* has a three-page reciprocal dedication to Dr Salter at the beginning, which clearly indicates that the two men got on extremely well together and that the doctor had been of great assistance to him. The dedication is included in the *Literature and Music* chapter.

Ernest, second from left, on the way to Svalbard. Source: David Booth, courtesy of Ross McNeill

Ernest in Svalbard
(1904–1913)

And now Mansfield is dead (except that I do not believe that any person really is dead), but Spitsbergen's future historian, if he is interested in any more than statistics, will count him in. And when our Arctic people gather, either in the forward cabin or the aft saloon, or round the oven in the barracks up there, Ernest Mansfield will be remembered with benevolence. A particularly better monument cannot be erected to anyone. (From Carl S. Sæther's obituary of Mansfield in the newspaper *Tromsø* 2.1.1935).

Svalbard loves its heroes and anti-heroes. They are characters who have popped up and disappeared again through various historical periods and in some way or other have left a mark that has become a story or a legend, often repeated, often with facts that become twisted as the tales get told and retold. There has never been an indigenous population in this high-Arctic archipelago, lying between 74° and 81°N. Before the international Treaty of Spitzbergen (Spitzbergen is the English and German spelling, Spitsbergen the Norwegian, now of the largest island in the Svalbard archipelago) was finally agreed upon in Paris in 1920, giving the sovereignty to Norway, Svalbard was No-Man's Land – or rather Every-Man's Land. Being first placed definitely on the maps after Willem Barentsz' visit in 1596, the freedom of access – weather and ice conditions permitting – brought individuals and groups of people from various countries to the islands to explore, to research, or – most usually – to exploit the natural resources. In the early seventeenth century it was summer hunting of whales and walruses, later followed by wintering hunters and trappers who were after polar bear and fox skins, as well as taking their share of the reindeer, walrus, seals and summer bird flocks, eggs and eider down.

Svalbard's coal reserves were known already to the Dutch and English whalers, who could fetch supplies from open seams to feed the train-oil boilers. Scientists and explorers could fill their ships' boilers with coal from the shorelines. In 1899 the first commercial load of coal was taken to Norway for sale. This was the start of what is known as the "Klondike period" in Svalbard, which lasted until the Treaty, and its associated Mining Code regulated the rights and responsibilities of mining ventures. Ernest Mansfield was one of the many who travelled to Svalbard in the early twentieth century and one of the minority who joined the exclusive gallery of Svalbard characters and legends.

Svalbard is often described as a geological picture book, where almost all the geological periods are represented and lie open for discovery without a camouflage of vegetation. Admittedly about 60% of the archipelago is covered by ice caps and glaciers, but the west coast of the largest and westernmost island, Spitsbergen, is tempered by a branch of a warmer current of water that originates in the Gulf Stream. It was along this more accessible, although mountainous, coast of Spitsbergen that the mineral prospecting took place during the first two decades of the 1900s. Prospectors from many countries including Norway, Sweden, Russia, Germany, Great Britain, the Netherlands and USA arrived each summer as soon as the ice allowed the ships to reach this Arctic Klondike and thrust their annexation signs into the permafrost to claim areas that they hoped contained economically viable reserves of coal, iron ore, zinc, asbestos, marble and even gold. There was no limit to the attempts and almost no limit to the areas that were claimed on more or less doubtful indications. The only system for staking claims was that signs should be erected that explained the extent, date and ownership and that this then be submitted to the claim-staker's own Foreign Office. The signs should preferably be renewed each year. There was no national body governing Svalbard and no international set of laws and rules.

How Ernest came to Svalbard

The story of how Ernest came to be involved in mining in Svalbard was recorded identically in at least seven newspapers in 1911 and also referred to in a colourful fashion in the *New York Times* of 13 August 1911. The original was an interview with Dr John Henry Salter in the UK *Daily Chronicle* (there were two *Chronicle* newspapers in the UK at this time: *The Daily Chronicle* of London and *The Essex Chronicle*) of 2 August and repeated in full in the New Zealand *Grey River Argus* of 27 September. This in turn was retold in the original bible of Svalbard's history, Adolf Hoel's three-volume *Svalbards historie 1596–1965*, and particularly from this latter source, has been spread into newer accounts of Ernest Mansfield's activities in Svalbard.

Adolf Hoel (1879–1964) was a Norwegian geologist who first visited Svalbard on a small topographical and geological mapping expedition in 1907 and who became totally fascinated with the beauty and majesty, and potential resources, of this unclaimed arctic archipelago. He devoted the rest of his life to what he maintained were natural Norwegian rights to the arctic areas from eastern Greenland to Franz Josef Land – "the Norwegian back garden", which he felt was the natural hunting ground for Norwegians who, after all, were born and raised in a land of ice and snow. He founded what became today's Norwegian Polar Institute and he was one of the major actors who gained sovereignty over Svalbard for Norway in 1920 (Barr 2003).

The Treaty of Spitzbergen is unique in the world in that it grants equal rights of economic opportunity to all citizens of nations which accede to the Treaty, and it is also an open treaty, which can still be acceded to today. However, during the discussions leading up to the Treaty in 1920 and its ratification in 1925, it was imperative for Hoel to ensure that as much territory as possible ended up in Norwegian and not foreign hands. Ernest Mansfield and the Northern Exploration Company (NEC) were therefore opponents that Hoel tried to deflate by various means – either by gathering witnesses who could state that the NEC claims were in fact previously claimed by Norwegian companies or by buying the NEC out of some areas. The accounts concerning Ernest and the NEC that we find in Hoel's publications are therefore coloured by his negative views of their activities in Svalbard and his attempts to play them down. This has in turn rubbed off into later accounts, and it is Hoel's negatively coloured and politically motivated description of Mansfield and his activities that has been a main component in the establishment of the popular version today of Mansfield's character.

Dr Salter's interview in the *Daily Chronicle*, 2 August 1911, as retold by Hoel does not differ greatly from the original, and the episode where the three pour over the rock specimens from Spitsbergen and Mansfield exclaims "Gold, or I'm a Dutchman"[1] is quoted in full in English. Hoel describes the men as three friends who had all travelled widely abroad. Dr Salter was an enthusiastic hunter who had fetched many trophies home from Siberia and northern Russia. Mansfield was one of the gold-mining pioneers of the Klondike and had also spent several years as a gold-miner in New Zealand, Australia and British Columbia. The Reverend Gardner was relatively wealthy and very travelled (see the chapter *Ernest in England* for details about Salter and Gardner).

Although Dr Salter was sceptical to many of Ernest's stories, the two men in fact

---

1 The phrase *It's Gold or I'm a Dutchman* appears in many places, but perhaps the most notable is in H. Rider Haggard's novels *A Tale of Three Lions* written in 1887 where Allan Quartermain and his son are prospecting for gold in the Transvaal, and in *Allan's Wife and other Tales*, also written in 1887. Perhaps Ernest Mansfield was an admirer of Rider Haggard.

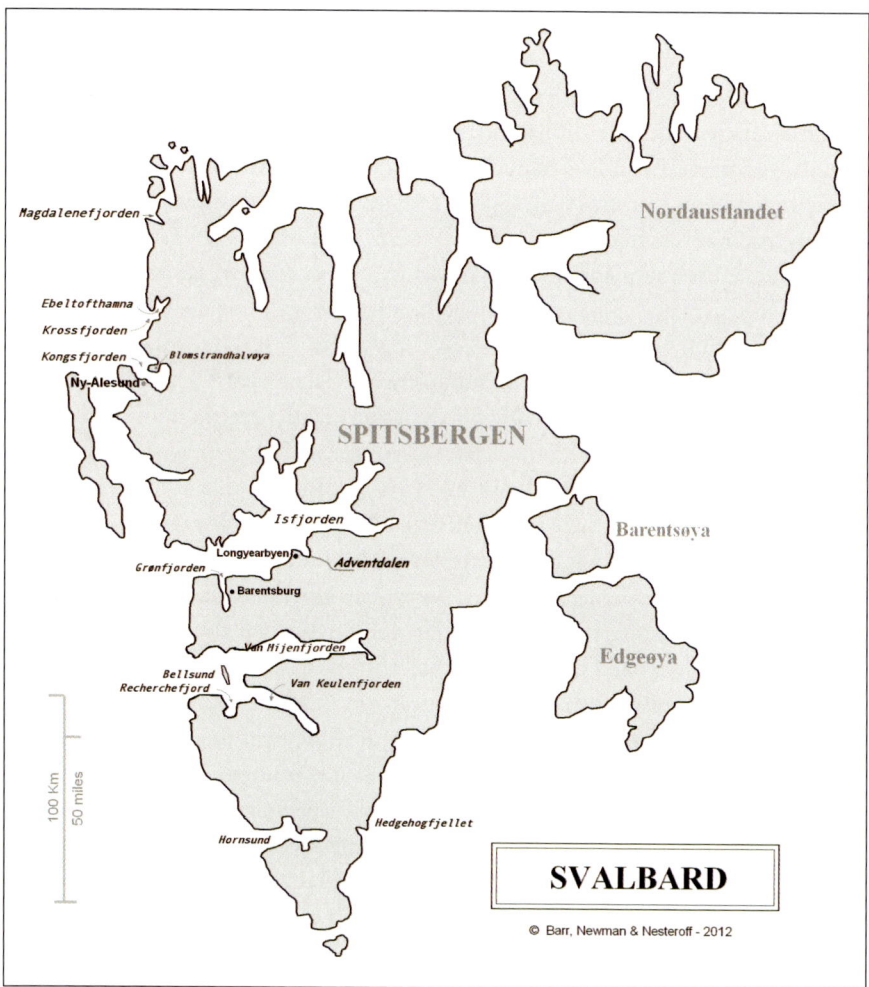

Produced by David Newman

became good friends. During their frequent conversations about various parts of the world, Ernest had particularly concentrated on gold and talked enthusiastically of his belief that the Arctic contained a large, and as yet undiscovered, wealth of the precious metal. Salter recalled the start of the Svalbard venture in spring 1904 in his *Diary of Reminiscences*; this incident is quoted fully in the *Ernest in England* chapter.

It is not certain exactly where the Rev. Gardner spent this first summer of 1904 in Svalbard, but he seems to have travelled on the tourist ship *Ophir* with his wife.

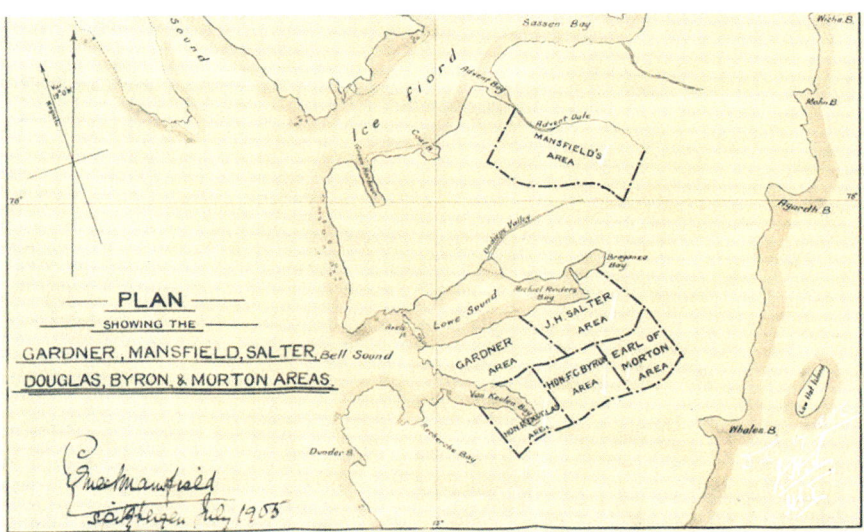

Ernest's signed claims map dated 1905. Source: the British Library

In a letter of 19 September 1904 to the British Foreign Minister, Lord Lansdowne, mentioned in Hoel's book, he asked for protection of the finds he had made and stated that they were on the west coast of Spitsbergen. It would seem that either Gardner, or more probably Ernest, was aware of the method of staking claims in No-Man's Land, i.e. that they should be reported to the claimer's foreign ministry as mentioned previously.

It is no surprise that the finds were made on "the west coast of Spitsbergen"; this was the most accessible area during the summer and the only area that had any general tourist cruise traffic at this time.

Ernest was now fired for a new venture in yet another area. Gold finds needed to be developed before anyone else could muscle in on them, and the following summer he and Gardner hired a small sloop – as mentioned in Salter's *Reminiscences* – *Familien* of Tromsø and sailed in the beginning of July the five days it took to reach Bellsund (Bell Sound) and Recherchefjorden (the Recherche Fjord). After some investigations ashore, *Familien* sailed on to Adventfjorden (the Advent Fjord), which today is the main settlement area and site of the Norwegian administrative centre and earlier coal mine Longyearbyen (originally Longyear City). In 1905 an American consortium, started by John Munroe Longyear (hence the name of the settlement), had already started a mine near the head of the fjord. Ernest and Gardner left their ship here. Gardner left for home on 17 July on a tourist ship, the German *Moltke*, while Ernest stayed until the 29th with engineer William D. Munroe who

Ernest in Svalbard | 79

led the American venture at Longyear City. He later left on another German cruise ship, *Prinzessin Victoria Louise*.

Swedish historian Dag Avango, who has researched the detailed history of the Swedish mining concern at Sveagruva (the Svea Mine), mentions that British Foreign Office archive material presents Gardner as the leading person in the venture on Spitsbergen with Mansfield as his engineer. In addition, he mentions that Mansfield after the 1905 expedition established the prospecting company *The Spitzbergen Mining and Exploration Syndicate*. His expedition in 1906 was organised through this company (Avango 2005).

The high Arctic "no man's archipelago" of Svalbard was by no means deserted at this time. Wintering trappers (mostly Norwegian) had their territories and small cabins spread around the archipelago, although the number of wintering trappers could vary; from 1905–1910 there were between 10 and 63 per year. The first all-year mining settlement was established in 1905 near Revneset by Adventfjorden and called Advent City. It was British and ceased activity in 1908. The American Arctic Coal Co. wintered in Longyear City for the first time in 1906–07, under the leadership of Englishman Bert Mangham. In addition, various companies had done summer prospecting along the mid-west coast since 1900. All these activities generated shipping traffic and increased the tourist cruise traffic, which had visited Svalbard already for a number of years. Land stations for whaling were established on Bjørnøya (Bear Island, the southernmost of the Svalbard islands) and at Finneset by Grønfjorden (the Green Fjord), near to today's Barentsburg. These closed in 1908 and 1912, respectively, but a Norwegian radio station was built at Finneset in 1911. Then there was also a significant traffic of hunting ships plying between Norway and Svalbard during the summer. It must be added that almost all the mining and shipping activity was concentrated along the west coast of Spitsbergen.

Adolf Hoel attempts to discredit claims

For Hoel it was important to ascertain whether Mansfield and Gardner had staked any claims that summer, and if so where they could have been. He concluded with satisfaction that they had not been in Kongsfjorden (the King's Fjord) – where there was later to be disagreement about NEC claims, and he obtained a testimony given under oath by the skipper of *Familien*, Jens Olsen, 2 May 1923 that the two Englishmen had been ashore no other places than the above-mentioned and that their trips on land had only been for hunting. Olsen had not noticed any signs or other means that they could have used to stake claims. Hoel added that back in England Mansfield was able to get his friends in Essex to contribute more capital to the venture so that a new expedition could be sent the following year. Olsen had previously sent Hoel a letter, dated 14 January 1919, stating "complete information" about Mansfield's and Gardner's visit to Svalbard in 1905. The two had not had others with them, and Olsen had regarded Mansfield as Gardner's servant. They had not visited other areas than Bellsund and Advent Bay. In further letters dated 15 and 18 February 1919 Olsen wrote that Mansfield "used serious lies" about the seaworthiness of Olsen's ship *Familien*, so it may be imagined that there had been some kind of disagreement between Mansfield and Olsen in 1905. This was perhaps the reason why both Mansfield and Gardner left the *Familien* at Adventfjorden and returned to Norway in other ships. Johan S. Hagerup, who met Mansfield in 1906 (see later) wrote to Hoel on 12 December 1918 that Mansfield had travelled only as a tourist in 1905, although it is not known how he gained that impression. Henrik B. Næss, a Norwegian captain of an Arctic hunting ship who also had interests in coal prospecting in Svalbard at the beginning of the 1900s, wrote to Hoel 20 March 1919 that Mansfield had stayed for about three weeks with Munroe at Longyear City and was completely broke. He was of the opinion that the tourists on the *Prinzessin Victoria Louise* had passed the hat round to pay for Mansfield's passage. Næss added that "He could really expound on his great plans, mostly I think about gold. I know nothing more about his travels around, since that man interests neither me nor Mr Munroe" (letter in Norsk Polarinstitutt archives).

Dr Salter had a different impression of the claims that Ernest and Gardner had made in 1905 as he wrote in his *Reminiscences*:

> They came back with certain parts of the country charted out, and to make our find secure we made our map ashore correspond with the cairns erected on the island, attaching to the portions of land the names of all the friends we had, so that we were really taking possession of the whole country. It was all well done. The gold and the coal were apportioned; there was also oil, and marble, and all sorts of other things. We thus felt that we had possession of something of world-wide value.

In 1906 the venture turned serious. Mansfield and Gardner took with them two others – Charles Mann and George Alexander (see *Ernest in England*) – and 20 Norwegian workers, travelling to Svalbard from Hammerfest, northernmost in Norway, in the steamship *Mylingen* in the latter part of June. Matters were hardening amongst the various mining companies, particularly in the Adventfjorden area, where claim jumping and quarrels about priority were the order of the day, and the American Arctic Coal Company (ACC)'s engineer Munroe would not allow any

Michelsen's cabin being renovated by Ernest and Charles Mann. Source: Charles Mann collection courtesy of Rosemary Mann

unloading of the *Mylingen* on the ACC claim area by Adventfjorden. A Sheffield company had recently claimed an area there that the ACC had already claimed the year before. Hoel notes that Mansfield took some of the Norwegian workers with him and walked some way up Adventdalen (the Advent valley is next to Longyearbyen). Since it was gold that was Mansfield's primary interest, Hoel reasoned that that was what he searched for in the valley. The expedition moved on to the next-door fjord, Sassenfjorden, and again Ernest went ashore with a few workers, but only in order to shoot some reindeer, according to Hoel. They then turned the ship southwards and visited Bellsund. A Bergen company had staked a claim for coal on the north shore of Van Mijenfjorden and erected a cabin ("the Michelsen cabin") there in 1901. It had been left stocked with various equipment, dynamite and other prospecting-related goods.

According to Hoel, Ernest and his men unloaded their own provisions and equipment here and placed them in the cabin. They then started to examine the coal layer between the cabin and the mountain Kolfjellet (the Coal Mountain). Two of the Norwegians, Nils Wasmuth and Oluf Martinsen, were left there as guards/caretakers at the cabin that Ernest now named Camp Morton after the Right Hon. Sholto George Watson Douglas, 19th Earl of Morton, Lord Dalkeith and Aberdour, who was also prospecting on Spitsbergen this summer. Rev. Gardner had, amongst other connections, been private chaplain to the Earl for eight weeks in 1899.

The rest of the expedition then steamed northwards to Kongsfjorden, and it was here that Hoel was eager to show that Mansfield had no claim to the south side of the fjord – in the area of today's settlement of Ny-Ålesund. Five years earlier a Bergen company had apparently claimed a large area where there were visible coal seams by stretching a wire from the shore towards the mountain, setting up a number of signs and erecting a cabin on the shore where they had stored tools and dynamite. As with the claim at Kolfjellet by Van Mijenfjorden, this had occurred in 1901. If no action had been carried out on the claims in the years since, it could be argued that the claims were no longer valid, and Hoel does not in fact accuse Mansfield of claim jumping in 1906 either in Van Mijenfjorden or Kongsfjorden. Exactly what had been done at Kongsfjorden that summer was crucial for Hoel's work between 1920 and 1925 to prove that claimed areas were Norwegian and not foreign. Ernest therefore gave a statement on 25 May 1922 to the Commissioner of Oaths in London about his activities in the Kongsfjorden area in summer 1906. He stated, as quoted by Hoel:

> I visited the area enclosed in red on the map now produced to me marked "E.M." and hereinafter referred to as the King's Bay [=Kongsfjorden] area, and in company with Charles Mann I carried a post to the top of Marble Island [today Blomstrandhalvøya = Blomstrand Peninsula] and affixed same with my name on it on the spot marked with a cross.
>
> I remained in this area in June and July and had a staff of about twenty men working in various parts of the North and South sides of the King's Bay area. With the help of eight or ten Norwegians I sank several pot holes and a shaft on the south shore of King's Bay the position of which is approximately shown on the said map and marked "S".
>
> In company with The Reverend F.T. Gardner and eight or ten Norwegians I pegged out the land by erecting posts and cairns with written claims and sardine tins under them claiming the whole of the land surrounding King's Bay and Cross Bay [neighbouring Krossfjorden] of a distance of ten miles from the water's edge and all the Islands in the Bays and near the shore.

Hoel could, however, quote a statement that Charles Mann gave on 11 April 1913 about the work at Kongsfjorden:

> At the end of June 1906 I accompanied Mr. Mansfield and Mr. George Alexander to Marble Island, Spitsbergen. We left Alexander in charge of the Norwegians and the boat and Mr. Mansfield and I landed on the Island and there erected a post and carved our names upon it and having done so returned to the boat.

Unloading supplies in 1906. Source: Charles Mann collection, courtesy of Rosemary Mann

Hoel found it worth noting that the lack of further details about the work at Kongsfjorden contrasted sharply with the details Mann gave about the work by Van Mijenfjorden that same summer. However, there was also a written statement given on 16 June 1919 in Tromsø by five of the Norwegian workers. They solemnly declared that they travelled with Mansfield in 1906 from Tromsø (Hoel – above – said that it was from Hammerfest) to Spitsbergen, where they worked coal seams and erected claim posts in several places. They sank holes and erected claim posts at many places both on the south and north side of Kongsfjorden as well as on the islands. They attested that they had seen no other people at any of the places where they had worked on Spitsbergen. It is interesting that Ernest's Norwegian workers supported him in this matter, even though they presumably only knew him from the work they had done with or for him in Svalbard. Nothing daunted, Hoel produced another witness with a more negative angle. Ice pilot Jørgen Mortensen witnessed in Tromsø on 22 January 1923 that Mansfield's ship had anchored on the south side of Kongsfjorden close to the cabin that had been erected in 1901. They had stayed by the cabin for eight days and dug "two small holes" in a search for the gold that Mansfield was sure was to be found there. During this digging, Mansfield with some of the workers rowed across the fjord to Marble Island and found the marble occurrences there. Mansfield took samples back with him.

Ernest at a claim post in 1911. Source: Marble Island prospectus, courtesy of Richard Gardner

Aerial photo of the wide Bellsund with Axeløya (Island) in the middle. Photo: Susan Barr

Hoel also found two Norwegian witnesses – Birger Jacobsen and Peder Engen – who signed a statement in Kristiania (Oslo) on 10 January 1919 that they had seen no claim posts belonging to Mansfield or the NEC in all the Kongsfjorden area in 1911 but had obliged by erecting claim signs backdated to 1905 in the mentioned areas in 1912 and 1913.

The compilation of these various witness accounts from different years enabled Hoel to draw his own conclusion that Mansfield's expedition had not staked any claims on the south side of Kongsfjorden. He reasoned that Mansfield was first and foremost out after gold and only came home with marble – from the other side of the fjord. He had not found gold and was obviously not interested in the coal on the south side since this was not investigated. Therefore, no claim was made, and thus the NEC in the early 1920s had no legal right to that area. Hoel covered all possibilities by stating that if Mansfield after all actually had staked a claim on the south side, he was too late anyway – with reference to the Norwegian company's claim in 1901. Many of Hoel's Norwegian witnesses can at the least be suspected of patriotic or dependency reasons for stating evidence against the English claims. Hoel's ideal of securing as much of Svalbard as possible for Norwegian ownership during the Treaty ratification period could have been shared by others. In addition, Hoel had become the hub of Svalbard activities in Norway, ranging from scientific expeditions to economic expeditions concerned with hunting, trapping and fishing, and it could be advantageous for hunters and skippers to keep on his side.

Ernest's expedition returned from Kongsfjorden to Camp Morton and unloaded all their equipment and provisions from *Mylingen*, which then returned to Norway. Charles Mann's statement of 11 April 1913 described what they then did. He, Mansfield, Gardner, Alexander and 13 Norwegians had been landed on the north side of Lowe Sound [Van Mijenfjorden] and had walked for 16 hours, staking claims all the way. They then went back to Camp Mansfield – which was the name given before Camp Morton – and claimed the area there, before going on to Van Keulenfjord and claiming the land on both sides. This is a huge area – and with no indications of minerals other than the previously claimed coal seams at Kolfjellet. This came to be one of the main arguments against Mansfield and the NEC – that large areas of western Svalbard were claimed without any substance regarding what could be mined or how. Mann went on to say that it had taken them about 10 weeks to do all this claiming. With the help of some of the Norwegians he also built Camp Mansfield – which included a blacksmith's shop with forge, fitted up the large cabin and timbered up the shafts of 50′ and 60′ which were driven into the coal seams. After the 10 weeks they left, leaving two Norwegians in charge of the camp for the winter. Odd Lønø, author of an overview of Norwegian trappers' winterings in Svalbard, has noted that the two men who wintered in "Mansfield's

Ernest (on the left) at the mine entrance. Source: Charles Mann collection, courtesy of Rosemary Mann

house where he had his coal and 'diamond' camp" at Bellsund were a part of Johan Hagerup's trapping expedition of 12 men spread in six cabins in the area.

While Ernest was at Camp Morton, Johan Hagerup arrived and declared himself to be caretaker for Michelsen's cabin from 1901. Hagerup (1846–1924) was a sealing skipper from Tromsø who also wintered several times in Svalbard on trapping expeditions. He protested against Mansfield's use of the cabin and the area, but, according to Hoel, the two quickly agreed that they would together pay for two trappers to spend the winter there and be paid for guarding the camp, while their catch of polar bear and fox skins would be divided between Mansfield and Hagerup. Mansfield's expedition was fetched to Norway by the steamer *Kjølva* in September. A short time later *Mylingen* returned to Camp Morton with six more winterers led by the Englishman Arthur Mangham, whose son Bert wintered at the same time as leader of the American ACC mining camp Longyear City.

1907

Mansfield's expedition to Svalbard in 1907 was of a less grand character than the previous year. He apparently travelled there by whaleboat and was landed at Rercherchefjorden. There he was later fetched in a "fairly frozen" condition by Hagerup and taken to Axeløya where Hagerup had spent the winter 1906–07. Mansfield had sent two Norwegians and some provisions up to Bellsund by another ship, *Thekla*, which returned to Norway with Hagerup and all the wintering Norwegians from Camp Morton. According to his own statement of 25 May 1922, Mansfield then revisited both shores of Kongsfjorden with 14 men and found all his claim signs still in place. Hoel was of the strong opinion that Mansfield here had mixed up his expeditions. From the information he gathered, he meant that Mansfield had no ship that summer that could have taken him north from Bellsund, and he had no more than six men with him. A wintering trapping expedition to Kongsfjorden 1908–09, led by Olaf Eriksen, declared that they had walked the south shore of Kongsfjorden almost daily and saw the claim signs and the cabin from 1901 but saw no signs that could have been erected either by Mansfield or the NEC. Hoel was not sure what work Mansfield's group had done in the Bellsund area that summer but presumed it was still gold he was searching for. The expedition returned to Norway in the autumn with ACC's ship *William D. Munroe*, while two or three Englishmen were, according to Hoel, left to winter at Camp Morton. Later that autumn Mansfield sent a ship up with more provisions for the winterers, but ice prevented it from getting to Bellsund.

According to the Goldhanger and Little Totham Parish Magazine of August 1907, the Rev. Gardner and his wife apparently left for a cruise to Svalbard in mid-July 1907, but the ship was turned back owing to bad ice conditions.

Camp Morton – photo taken by David Booth in 1912. Courtesy of Ross McNeill

Ernest winters 1908–09 by Bellsund

For the summer expedition in 1908 Mansfield hired the stream-driven fishing boat *Kvædfjord*, which took them to Bellsund. In his statement of 11 April 1913, Charles Mann declared that he built Camp Bell this summer, which took seven weeks and could house 16 men. Mansfield put his men to work to collect rock specimens which he hoped would show signs of platinum and gold. *Kvædfjord* took the specimens and most of the men to Norway and left Mansfield and four men. Dr Salter noted in his diary for 20 August 1908: "A telegram that the Spitzbergen assay was 16 grains per ton and therefore "useless"". It was decided that Mansfield and his group should winter at Camp Bell, and *Kjølva* brought provisions for them in the autumn. Mansfield stated that he was at Kongsfjorden again this summer and still found his claim signs intact and no other persons in the area. He was accompanied by five Norwegians – Nils Wasmuth, Ole Martinsen, Hans Norberg, Hans Nilsen and Alfred Johannesen – and several others. Hoel reckoned that this applied only to the Blomstrand area (Marble Island). In early summer 1909 Johan Hagerup arrived in the *Bellsund* to relieve his two trappers in the Bellsund area. He also had post for Mansfield and went therefore to Camp Bell. Mansfield asked for a passage back to Norway for himself and his five men, and in August he was back in England.

The period described above covers an interesting occurrence – that Mansfield, an English mineral prospector and not, for example, a seasoned hunter, should choose to spend a long, dark and potentially hazardous winter in a small cabin in such an isolated high arctic area, and with no "soul mates" as company. There was little chance of mak-

Ernest in his Camp Morton cabin. Charles Mann collection, courtesy of Rosemary Mann

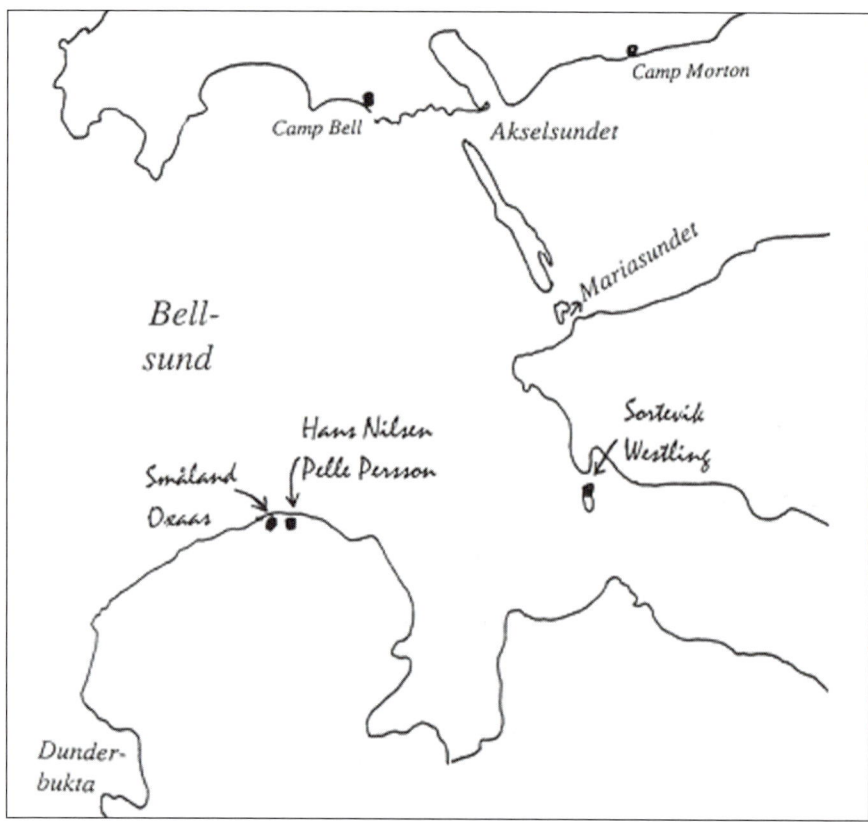

Wintering cabins around Bellsund 1908–09 (Map from Lønø 1994:123)

ing the strike of the century during that winter, i.e.: perhaps finding the elusive gold vein. So why submit himself to this experience? Perhaps it was his romantic nature coming to the fore – a desire to feel in the strongest way possible this area of the world to which he was dedicating his summers. It resulted at the very least in his first novel, the romantic and science-fiction-like *Astria – The Ice Maiden*, published in 1910, with a storyline relating very strongly to Ernest's earlier life and to his experience of wintering in the Arctic wilderness (see the chapter *Ernest's literary endeavours*).

There are several other accounts of the winter 1908–09 by Bellsund. Arthur Oxaas, another of the historical Svalbard personalities after his many years as wintering hunter and trapper, spent that same winter at Calypsostranda by Recherchefjorden. On the north side of Bellsund, Oluf Olsen and Nils Wasmuth wintered in Camp Morton for Mansfield, while he himself was in Camp Bell with Hans Norberg. Around Bellsund this winter, there were in fact ten men living in five cabins.

In his autobiography from 1955, Oxaas wrote that he and two others visited Camp Morton in February to chat with Olsen and Wasmuth but found only a note left there stating that they had gone to Camp Bell to celebrate Christmas with Mr Mansfield. Oxaas wrote dryly "It was certainly a long Christmas celebration". Olav Sortevik and Kalle Wesling on Eholmen (Eider Islet) developed scurvy in March and were first helped to the cabin next to Oxaas', and then in May Oxaas and two others made their way by boat to Mansfield on the other side of the fjord to get help. Oxaas wrote in his account that "At Mr Mansfield's we were received with open arms. He gave us a large supply of dried potatoes, lime juice, condensed milk and dried fruit. Altogether he was willing to do whatever he could to help the sick". Unfortunately Olav Sortevik died. Oxaas' book otherwise contains a very sympathetic description of Mansfield and of the Northern Exploration Company, which he gained from connections with the early prospecting expeditions.

Odd Lønø states in his account of Norwegian trappers' winterings that Mansfield's aim was to prospect for coal, but he also took samples of stone and gravel that he thought contained gold and platinum. Lønø adds that Mansfield had given orders that he should take the first shot at any polar bears. "They saw some bears, and Mansfield shot over and under to scare them away". Lønø cites Hoel, Oxaas and Arvid Moberg (author of *Svalbards sønner*, Eides forlag 1960) as his sources of information.

Ernest's wintering in 1908–09 was described by Dr Salter in his *Reminiscences* with a touch of the dramatics that Ernest himself used in his novel *Astria – The Ice Maiden*:

> Mansfield must have had a dreadful time, for he remained in Spitzbergen alone the whole of one winter. There was some reason why it was important that he should do so, and I do not think he could get his Norwegians to stay with him. He was there alone. There were awful storms, and he built himself a wooden shanty high up on the mountainside – of course on a mountain close to the sea, but apparently out of the way of any rising of the waters. The mountains get very quickly out of the sea and ascend to some heights. The water rises to abnormal heights, and on one particular night great icebergs came floating about, and, governed by eddies of air, began cannoning one another in the ocean. Mansfield heard a tremendous cannonade going on, and looking out saw nothing but icebergs round him, sailing about and gambolling together, making a diabolical noise. He thought his house would be annihilated, in which case it would not be safe for him to be inside, so he wrapped himself up in blankets and went and sat in a cave in the mountain – a long time, bitterly cold – and watched this storm of Nature such as he had never seen or conceived before.
>
> "There I was, all alone," he told me, "all those months, and the beautiful little silver foxes seemed only to require a bit of enticing to come in. They came in, and were just like our tame dogs at home. They would sit up and beg for food, and take their food in that way" – and he showed me photographs in proof. The skins of these silver foxes were worth about £40 apiece, and the strange thing was that as soon as the Norwegians came in the spring they sloped. Close to Mansfield's house one morning there were three polar bears, and he brought a skin of a huge bear home for me.

It would seem that Ernest did not only shoot to scare the bears away, or maybe it was one of the trappers who supplied the skin.

The NEC is founded

Ernest was back in Svalbard in mid-August 1910 with the *Aurora*. Some men who were to work and winter at Camp Bell under the leadership of Arthur Mangham were put ashore there. According to his statement, Ernest spent the summer at Kongsfjorden and had the cabin Camp Zoe built this year. Ernest's daughter Bernice Zoe had been born in 1904.

Adolf Hoel could, however, quote another source who maintained that the cabin was not built before September 1911 – this is borne out by the accounts here by the Norwegian trappers who wintered for Ernest in the Kongsfjorden/Krossfjorden area – and again that Mansfield had only stayed for a short while by Kongsfjorden before returning to Bellsund and Van Mijenfjorden. He visited Braganzavågen in that area – where the Norwegian mining settlement Sveagruva now lies – and had the cabin Camp Williamson erected. (This cabin no longer exists). Henry Williamson had contributed to the expedition and later became a

Ernest with his wife and daughter wearing fox skins. Source: David Booth, courtesy of Ross McNeill

Mining remains at Camp Millar 1998. Photo: Susan Barr

board member of The Northern Exploration Company Ltd. Three cabins were also erected on the north side of Bellsund, by Ingeborgfjellet, east of Camp Bell. Ernest called this Camp Millar, again after one of the contributors to the venture who later became one of the biggest investors in the NEC.

Four Norwegian trappers were to winter at Braganzavågen for Ernest this winter 1910–11. On top of this both the *Aurora* and two summer hunting expedition ships, the Tromsø ship *Hvidfisken* and *Prøven* of Vardø were caught in the ice in Recherchefjorden in September, and it looked as though the three crews were in for a long and hungry winter. The crews of *Prøven* and *Hvidfisken* were able to make their way over the ice to Camp Bell where Mangham gave them a sack of flour, half a sack of biscuit, and a little butter and sugar. At the end of September *Aurora* struggled out from Van Mijenfjorden to reach Camp Bell and heard there about the other two ships. The skipper took *Aurora* to Longyear City to stock up with supplies and coal and steamed back to Bellsund to rescue the other two crews from the unplanned wintering. The two ships were left frozen in each with one man on board, who volunteered to stay for the winter in order both to hunt and to look after the ships.

By this time Ernest's expeditions to Svalbard had cost £17 400 (c. £1.7 mill. today). No gold had been found, but the marble occurrences at Marble Island seemed well worth further investment. Nathan H. Dole in his account from 1922 of the American Arctic Coal Company recounts that William Scoresby jr (1789–1857) c.100 years earlier had remarked on the marble by Kongsfjorden as being "of real beauty". It was decided that a limited company should be formed. Hoel names Horace Montague Hobrow as the initiator. Hobrow was a solicitor of the High Court and partner in the law firm Rundle & Hobrow of London. The Northern Exploration Company Ltd. (NEC) was registered on 15 November 1910, with founders F. W. Hobrow and Vernon F. Williams of London, and Charles W. Edwards of Isleworth. Each of these owned one share à £1 of a total of 100 shares. The Company's activities were not planned to be limited to Svalbard, and according to the Statutes they were to be extended to prospecting and mining of all "minerals, precious stones, and oil... in any part of the world". The primary task was, however, to acquire more capital, and on 3 April 1911 a further 124 900 shares at £1 were issued. The number of board members was increased to five and the following were appointed: H. E Millar of London (chairman), Ernest Williamson of London (adm. dir.), Sidney Thomas Peirson (secretary), Gerald Dudley Smith of London and Fred. Lewis Davis of Potters Bar, Hertfordshire. On 2 May 1911 the Board informed the Companies' Registration Office in London that the Company had taken over from Ernest Mansfield all his claims and rights of all kinds on Spitsbergen for the sum of £75 000. The amount of £5000 was to be paid six months after the signing of the contract on 23 May, while the remainder would be paid with £1 shares in the Company. In addition, Mansfield committed himself to

Marble Island/Ny-London. Source: David Booth, courtesy of Ross McNeill

Ny-London in 2008. Photo: Susan Barr

working for the Company for five years with an annual salary of £500 (Land Sale Contract in the official Register of Companies in London). According to addresses used in NEC papers, the Company was officially based in the City of London.

Dag Avango in his history of Sveagruva reveals from British Foreign Office documents that Mansfield brought to the foundation of the Northern Exploration Company claims from a number of other names than only his own Spitzbergen Mining and Exploration Company. These were "The Honourable Roderick Douglas claim, the Earl of Morton claim, the Reverend F. C. Byron claim, the J. H. Salter claim, the Countess of Morton claim and D. Campbell and the Honourable Edwin Ponsonby claim". Since most of these names are known to have been associated with Ernest's activities on Spitsbergen, it may be presumed that his financial backers were apportioned claims relating to their contributions.

On 7 July 1911 the Company reported to the Register that Fred. Lewis Davis had taken over 5000 shares, H. M. Hobrow 500 and Gerald Dudley Smith 1000, each paying 10 shillings a share. A week later another 25 000 shares of 10 shillings each were subscribed for by 27 shareholders. Of Ernest's 70 000 shares, he turned over 6400 to the Rev. Gardner, 6500 to Dr Salter, 31 750 to Henry Williamson and 18 250 to the chairman of the board H. E. Millar. In the 1911/12 Prospectus the Rev. Gardner wrote: "all the interests of myself and friends in 'N' have been legally transferred to Mr EM, our Representative" ["N" being the Prospectus euphemism for Spitsbergen]. So presumably, Ernest was repaying them, which is an indication that they all got on extremely well and trusted each other.

The 1911 Expedition

In 1911 the Company sent a large expedition to Svalbard, led by Ernest, with the English steamship *Repetor* (Oxaas writes *Repertoir*). Ernest had hired about 20 quarrymen in Aberdeen to mine the marble at Marble Island, and from Norway he brought several carpenters who were to erect more buildings. The Scottish connection is, incidentally, interesting. Ernest's second wife was Scottish, the Scottish Earl of Morton was involved in the NEC and Scottish workers and associates were often brought on to the expeditions. Heavy machinery, including winches, cranes, materials for a short rail track and building materials were brought from England. The landing site at Blomstrandhalvøya was named Peirson Harbour after the Company secretary, and during the summer a smithy, storage buildings and two living quarters were erected and a rail track laid from the marble quarry to the harbour. Another cabin was erected on Storholmen, an island in the Kongsfjorden, and named Camp Davis.

The result for the summer was that 50 tons of marble were quarried and shipped as specimens to England. On 20 August five of the shareholders, together with Peirson and Williamson, arrived at the camp, where they spent five days inspecting the workings. They could report with satisfaction back to the company that the island was indeed composed of marble. Ernest then joined them on the *Repetor* to travel south to Camp Bell and inspect the workings there, before they returned to Norway at the end of the month. Hoel quotes Mansfield's statement where he declared that the group – Englishmen and Norwegians – also visited the south shore of Kongsfjorden and noted that the claim signs and the mine shaft from 1906 were still in place. As usual, Hoel then produced a witness, this time the Norwegian Jacob Arnesen, who could swear under oath in Tromsø on 6 November 1922 that Mansfield's expedition at no time that summer landed on the south shore.

Hans Larsen Norberg (1861–1917) participated in several Norwegian and Swedish geological expeditions to Svalbard, as well as Mansfield's, and was praised in his obituary by Swedish professor A. G. Nathorst for having on the basis of his extensive experience "made important contributions to the knowledge of this land's [Svalbard's] nature and history". Norberg sent his observations on the geology of the Bellsund area to Nathorst and participated in Swedish coal prospect-

Marble Island/Ny-London with the loading crane in the foreground and Port Peirson behind. Photo: Susan Barr 2008

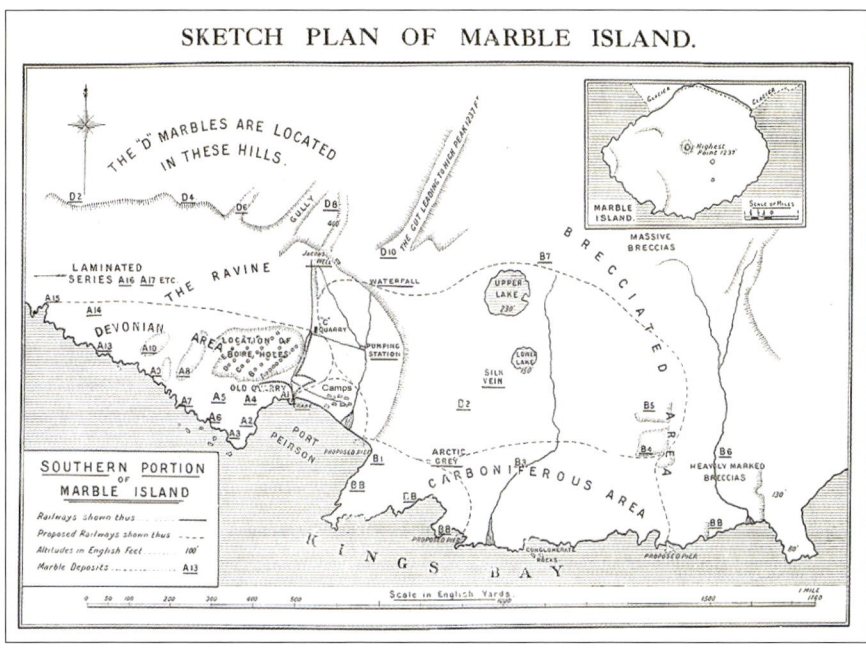

Source: Marble Island prospectus, courtesy of Richard Gardner

ing there in 1911. On arriving in the Bellsund area, he informed the Swedes of Mansfield's investigations and occupations in 1906 and expressed the opinion that Mansfield would never come back to Svalbard. It was only when the expedition returned to Sweden that Norberg learned from Norwegian newspapers that Mansfield was returning with a new expedition that same autumn. Swedish interests opened the Sveagruva coal mine in the area in 1917. It was sold to Norway in 1934 and is now the most productive mine in Svalbard. It may be said that Mansfield's expedition in 1906 was the first to occupy the area on the basis of coal occurrences. They were, however, not the first to note the coal there. Dag Avango notes that a merchant named Ivar Stenhjelm from Vardø together with two others from Bergen discovered in 1901 the coal occurrences in the Bellsund-Van Mijenfjorden-Braganzavågen area that Ernest was to occupy five years later. It was this expedition that erected the Michelsen cabin on the north shore of Bellsund.

American John Munroe Longyear had started up The Arctic Coal Company (ACC) and the Longyear City mining settlement (today's Longyearbyen) in 1906. In summer 1911 the miners went on strike and the two ringleaders precipitated a violent situation, which the manager, John Gibson, managed to get under control. Longyear visited Svalbard again that summer and on his return to Tromsø he was contacted by Mansfield who had heard of the troubles in Longyear City. Mansfield suggested that the ACC and NEC join together in "an offensive and defensive alliance for mutual protection in case of insurrections, strikes or riots", but Longyear declined (Hoel:642–43 and Dole 1922 Vol.II:91).

Nathan Haskell Dole wrote the history of Longyear's coal mine at Longyear's request. Called *America in Spitsbergen, the romance of an Arctic coal mine*, it was published in 1922, the year Longyear died. Dole states (p.106–07) that Mansfield had apparently been in America in 1911 to buy machinery. Dole wrote disparagingly of "interloping claimants … pushing their audacities" about companies such as the NEC. Later, however, (p.226) he describes a visit that Longyear made to the NEC camp at Blomstrand where he was impressed by the orderliness and friendliness there. The camp consisted of about a dozen buildings, Dole quoted from Longyear's report, of which eight were dwellings. There was also a large warehouse covered with galvanised iron, and a considerable railroad of 5 ft gauge was being constructed. Longyear was shown everything both at the mining camp and the area around and he had no criticism to convey. The difference in opinion here can of course be that the ACC had no conflicting claims with the NEC in this area as they did in the Isfjorden area.

In September 1911 a wooden house that consul Johannes H. Giæver had erected by Recherchefjorden for use by tourists, together with an occupation he had made in the area, was sold to Mansfield for £300 (NOK 5400). That autumn Ernest

Giæver's house by Recherchefjorden in 1998. Photo: Susan Barr

engaged trapper Arthur Oxaas to stay as winter caretaker 1911–12 at Blomstrandhalvøya together with three men that *Repetor* fetched from Tromsø. A small caretaker group was also landed at Camp Millar, Bellsund, where they were to drive an adit into the hillside under Mangham's leadership. At "London" on Blomstrandhalvøya, the about 60 English and Scottish marble workers were waiting to return home, since the production had not turned out to be economical. The mine working was concluded and the workers sent home, while Oxaas, his brother, August Stenersen and the Scot James Butt moved into one of the large housing barracks. "James Butt" whom Oxaas mentions in his book was most probably Mansfield's brother-in-law James Booth, who wintered at the marble camp in 1912–13 (see later). Two other Norwegian trappers, August Olafson (the name also appears as Olafsson and Olofsson) and Henry Rudi, were sent to Krossfjorden where the cabin Camp Zoe was erected to house them.

Lønø states that the wintering trappers Mansfield brought from Tromsø were to receive kr 50 per month (c. NOK 2720 in 2011), free supplies of food and tobacco and half the skins they trapped. "They had never heard the likes before for winterers. No rationing or anything. They would even get stoves to take with them so that they could bake bread. No more chewing on unleavened pancake bread. And finally, the best trapping team would get a gold watch!" (Lønø 1998:93). The two trappers who were left at Recherchefjorden, Gustav Lindquist and Ed-

One of the marble quarries on Blomstrandhalvøya. Photo: Susan Barr

vard Abrahamsen, had the company of some miners who were left there. Early in May Lindquist walked to Grønfjorden to fetch the doctor to a sick Englishman at Camp Millar, but the man died the day after the doctor arrived.

Oxaas witnessed in Tromsø on 16 January 1923 that his instructions for the winter were to look after the property and claim signs on the north side of Kongsfjorden and the islands in the fjord. No mention was made of the south side. Oxaas also stated that several claims signs around Blomstrandhalvøya were put up in 1911, but marked 1905, and that other signs marked 1905 had appeared in Krossfjorden in 1915 that were not there in 1911 (Hoel 434–35 and Oxaas).

Oxaas relates in his book the well-known story of Mansfield's debut in Svalbard, after he had gathered mining experience abroad, for example in British Columbia; how he met Gardner and travelled with him to Spitsbergen; how the marble occurrences at Blomstrandhalvøya had seemed so promising, but that "the disappointment was indescribable when it turned out that, after standing for a while in the sun, the marble blocks crumbled up. The stone was frozen and became completely shattered in the warmth". He stated that they tried drilling to

100 m depth, but the result was the same and the activity therefore doomed. At the end of the War, they tried again with prime mover Salisbury Jones. He was able to gain £125 000 in new share capital, but the new prospecting attempts only produced disappointing results.

Oxaas then described his own dealings with the NEC. He met Mansfield by chance in Tromsø in 1911, after having last seen him in Bellsund in 1908. Mansfield was looking for trappers to hire as caretakers at the NEC mines for the winter. Oxaas was hired as leader at one of the stations and after a week gathering men and supplies, Mansfield sent Oxaas and the others in the *SS Repertoir* to Svalbard. Henry Rudi was one of the other trappers, and like Oxaas he became a famous Svalbard personality through his book describing his many years as a trapper in Svalbard. The expedition stopped first at Camp Millar in Bellsund and also visited Braganza and Sørfjorden to leave caretakers and then arrived at Kongsfjorden and Marble Island, "or 'London' as this marble town was called".

> Here a number of houses had been erected, so the place was unrecognisable. There were living houses and storage huts and from the quay a railway led up to the adits in the hillside where the marble was taken out. There was even a locomotive. About 60 English and Scottish marble workers waited there to be taken home since the working, as I have previously mentioned, was not economic. The strange thing with this little town that had grown up here was that each house had an English person name. They were simply named after the most important of the Company's shareholders and other famous people.
>
> When the work in the marble town was finished and the workers were sent home, August Stenersen, my brother and I moved ashore and into one of the large living houses, which by the way was called Camp Person. Trappers Olafson and Henry Rudi were sent into Krossfjorden, where a little house had been built and named after Mr Mansfield's daughter Sue. Enough about that, as the fourth man in our team we got a Scot called James Butt. We four were now to look after the place and erect claim signs around the fjord. In addition we were supposed to hunt as far as we had time to do so. The house we lived in was large and grand, and there was plenty of food. Here we could revel in all kinds of lovely things, such as dried fruits and tinned fruit. And the supplies were so plentiful that we didn't need to offer rationing a thought. There was assuredly food here for several years on. In addition we had with us salt meat and fish as well as dried fish from Tromsø. The store houses were also richly supplied with tobacco and cigars and for pastimes we had all kinds of games to choose between. A phonograph was not missing either. It is doubtful that any Norwegian trapper had ever experienced the likes of such wintering luxuries before.

In fact, all the trappers who wintered as caretakers for Mansfield had only positive opinions about his generous provisions and his consideration for them.

To start with we mainly worked to get through the marking work. We were around in the terrain along both sides of the fjord and erected claim signs, which were rectangular metal plates with the Company's name and the year on. The year was incidentally old. It showed several years previous to 1911.

Later in the autumn, when we were to attempt some trapping, we divided into two groups. August Stenersen and my brother went to a cabin on Kapp Quise [Guissez], while the Scot and I stayed at home. It was James Butt's first wintering. In fact he had never been away from home before, so he was not free from homesickness. Each evening we wound up the phonograph and could enjoy a rich music programme. Absolutely something for every taste. We could choose between everything possible from Hawaiian music to the finest orchestra concerts. But James had in fact only one favourite melody and that was called just "My Mother". He played that every evening, and there were tears in his eyes (Oxaas 1955).

An NEC claim sign. Source: Marble Island prospectus, courtesy of Richard Gardner

Oxaas described Butt as a pleasant and easy comrade and they soon learned to understand each other's language. James was clever and eager to learn. He had never been in a boat before but soon learned to row, and in winter to ski and hunt, never complaining when the skiing proved almost too strenuous. At Christmas they all joined together and had the "world's finest supplies" to choose from. While James listened to *My Mother*, the others played casino and dominos. There was also plenty of reading matter for all. They shot a few foxes but saw no polar bears. After New Year Olafson and Henry Rudi came visiting and reported on good foxhunting in Krossfjorden. In February fulmar started arriving, even though the temperature was minus 25–30°C, and the men could start shooting for fresh meat, and later there were seal pups to add to the menu and sealskins to collect to add to their income. As the spring advanced Oxaas and Stenersen travelled with the boat to collect eggs and eiderdown, while James and Oxaas' brother were to get Camp Peirson ready for the arrival of Mr Mansfield or others from the NEC. Oxaas and Stenersen travelled as far as Grønfjorden, where they met Mansfield at the beginning of July travelling north on the *Activ*.

Cattle provided for fresh meat. Source: Marble Island prospectus, courtesy of Richard Gardner

According to the contract, Mansfield had the right to half of the eiderdown the men had collected that spring, but according to Oxaas he was "too generous to bother with such things, - Keep it all, he said and threw out his hands. Mr Mansfield was no petty man. He wanted people in his service to be satisfied, and he did not count the small change when it came to payment". The watchmen travelled back in July to Tromsø with the *Munroe*.

Rudi and Olafson had been put ashore at Tinayrebukta for the winter 1911–12 with building materials for a cabin. They erected a fine cabin and Mansfield asked whether they could name it after his only child, Zoe. It was called Camp Zoe. As the winter set in with its trapping activities, Rudi moved to a cabin at Kapp Mitra, a little to the west. They saw no polar bears during the winter. Mansfield arrived at Kapp Mitra in the middle of June. He had collected all the caretakers on the ship and he now paid out for the work and the winter catch. Each group visited him in his cabin and reported on their activities. Again Mansfield showed his generous nature; he paid up to kr 20 more for each fox skin than they would have fetched in Tromsø. The trappers could hardly believe it. Since Olafson and Rudi had the biggest catch, they each received a gold watch plus kr 100 extra. As Lønø remarks: "Unbelievable"! Rudi told in his book (Rudi 1958) that there had also been a group of two Scotsmen and a Norwegian, and they had only caught one fox during the winter. Mansfield paid for the skin and then let them keep it.

Not unexpectedly, there was a celebration when the trappers were delivered in Tromsø by *Repetor*. Lønø quotes Rudi's description: "We turned up at the brewery

The NEC cook with a bird for the pot. Source: Ross McNeill

and bought cash down one of the large storage barrels of beer, it took at least 500 litres. Then we haggled over the hire of the brewery's strongest horse, it was to pull the beer to the south end of Tromsøya for us. At the same time we hired two buckets and six ladles … Then we rushed to town and bought a couple of large smoked salmon, some cups and such like, and then we started to invite guests …".
It was not unusual for the trappers from Svalbard to blow off any profit they might have after a long wintering on the joys of being back in town with some money to jingle in the pocket.

Another Norwegian trapper who wintered for Mansfield 1911–12 was Torberg Pedersen. According to his diary (NP), he wintered at Camp Margaret by Braganza Bay (Braganzavågen: where Sveagruva is today) with Englishman Charles Hayward and two Norwegians, Julius Jensen and Thorolf Olsen. The agreement was as usual that they would hunt and trap as well as take care of the NEC property. They were left with first-class provisions in abundance. Their cabin lay across Braganzavågen from Camp Williamson and about 80 km from Camp Bell. According to Pedersen the Englishman, Charles Hayward, turned out to be "intolerable – shoddy and dirty, but I suppose we should excuse him – he is after all just an Englishman and says he apparently is a shareholder in the company". Pedersen was told by other winterers that Mansfield's cabins were grand in comparison with those many other leaders offered. There was even a gramophone. In March four of the men from Camp Millar arrived to see whether the Swedes who had prospected in the area that summer had left any dynamite behind, since there was none left at Camp Millar. They could not, however, get to the Swedish mine owing to the large amount of snow in the area. On the way to Camp Margaret, they had had to leave behind an Englishman who was too exhausted to continue. "Luckily this was not more than c. 10 km from our cabin so we were able during the night to find the unfortunate man and bring him back, almost unconscious. Here you have immediate proof that Englishmen cannot measure up to us Norwegians in strength and stamina, but luckily we saved his life and that was the main thing". Pedersen exchanged with one of the men a few days later and walked back with the others to Camp Millar to work in the mine there. On the way they passed Camp Morton "where Mansfield some time back had had a trial mine for coal". During the winter they worked on the mine as planned, but coal for fuel was fetched from a face 7 km away. In addition, they collected driftwood. According to Pedersen, during the winter the workers dynamited out approximately 1000 tons of stone, which was removed from the mine with the help of a crane. They reckoned that the gold-quartz or gold occurrence became richer the further down they reached. Pedersen described Mangham as being a nice person to talk with but not a good mining leader because he fired two men who complained that it was too cold to stand at the mine while a north-easterly storm was blowing. He also complained that there was neither a doctor available nor enough medical equipment and not even bedclothes for a visiting doctor. The fault was Mangham's according to Pedersen.

Pedersen had a great belief in Svalbard's mineral wealth and meant that they had found gold; shortly after in his diary he modifies this to "possible traces of gold", but with a good occurrence of coal and anthracite. Five men were working every day both in the coal and the gold workings, but no proper vein had been found. Towards the end of June, Pedersen reported a dismal atmosphere among the men. They had run out of butter and Mansfield had not appeared. He had promised to be there at the end of May but had still not arrived at the end of June. This was about the same time that Mansfield and his summer expedition were leaving London (22 June, see later) and it was by no means unusual that wintering trappers and others in Svalbard had to wait until mid-summer before ships could get through the ice to relieve them. Nor was it unusual for certain attractive supplies to run out before the end of the season. However, Pedersen reasoned on in his diary, Mangham had the responsibility for 11 workmen who now could not work in the "gold mine" (Pedersen's quotation marks) because of water seeping in, there was no coal in the mountain where they thought it would be, the supplies were almost used up and he meant that Mangham could have done something and not just waited for Mansfield. With no radio communication, the only action possible would have been for someone to make their way overland c. 50 km to the new telegraph station at Finneset by Grønfjorden (Spitsbergen Radio telegraph station was established at Finneset, near today's settlement of Barentsburg, in 1911), which Pedersen and one other had done in June in order to collect telegraphed post. In early July he was sent on a passing ship belonging to a Norwegian expedition (Arve Staxrud's) back to Grønfjorden to fetch some supplies and arrived back with another boat just before Mansfield finally arrived on 11 July, after apparently having been held up in London for five weeks owing to a dock strike. New supplies were offloaded and Pedersen, another Norwegian and an Englishman were left at the site as watchmen, while Mansfield took 52 workmen to Blomstrandhalvøya, where he was also to land six houses (this does not quite match the schedule given below). In addition, they were to spring-clean the buildings at Camp Millar and dig drainage ditches around the houses. Several more houses were to be built at Camp Millar as there were to be 16 men there during the coming winter, during which they would also mine zinc. Pedersen returned to Norway at the end of the summer with a reference from Mangham stating that Pedersen had worked for him since 19 March 1912 and had shown himself to be a good worker "at all kinds of work and willing to do the work he is set to".

In April 1912 H. E. Millar resigned from the Board because of illness and he died 12 July that year. His place was taken by Oscar Emanuel Warburg of London (Hoel:435–36). Also in April, a Norwegian speculator – Birger Jacobsen – sold various claims he had around Bellsund, at Hornsund and by Grønfjorden to the NEC and went to work for the Company as a prospector. The claims had been made for coal, galena, asbestos, nickel and gold. Jacobsen had published an article in the Norwegian newspaper *Verdens Gang* 30 September 1911 entitled *Det lovløse Spitsbergen – Fra et videnskabelig og økonomisk synspunkt* (Lawless Spitsbergen – From a scientific and economic point of view). Here he urged the Norwegian authorities to start to take responsibility for the Norwegian claims. Everything was plundered by Norwegian trappers, for example "Mr Mansfield's house on the coast by Axeløya in

Bellsund was, during my last visit in 1909, well equipped with supplies of all kinds. There was also a notice stating that those who needed any necessities could help themselves. That same autumn and winter the house was simply robbed for equipment and supplies. Norwegian winterers plundered lock, stock and barrel. Rumours even say that large amounts were taken home to Norway".

Birger Jacobsen related in some autobiographical notes now in the NP archives (dated when transcribed in 1935 from his hand-written notes) that he worked on Spitsbergen every summer from 1912 to 1921 with Mansfield as manager, apart from two years during World War I. He described the NEC as a pronounced "landholding company" which occupied as much land as possible without consideration for whether or not there were any viable mineral resources in the area. He stated that the Company's leadership had little or no belief in economically viable coal mining, and Mansfield's only real objective was therefore to find gold. His earlier years of experience in "Alaska, South Africa and Australia" had given him the taste for this valuable but rare metal. However, Jacobsen added, "where there is nothing, even the Emperor has lost his rights". He tells a story of Mr Maugham (actually Mangham), the English shift leader in Bellsund winter 1911–12 who was eager to drive a shaft into what they all thought was gold, but which turned out to be silicate, proving the fact that "All that glistens is not gold". A telegram had been sent as soon as was possible to London reporting "Plenty of gold, Expect instructions" and three barrels of the mineral were shipped to London at the first opportunity. There was no little disappointment in the board room in Fenchurch Street when the "gold" was immediately recognised by the managing director as iron pyrites. According to Jacobsen "Maugham gold" from that day on was the name used by the Company's employees for all glittering stone from pyrites to graphite.

Dr Salter recalls in his *Reminiscences* that:

> The "Captain" of our mine, in charge of our works there, was Maugham, a very reliable man who did not speak much, but on his way home on one occasion he happened to let fall a word on board ship, which word was carried to somebody else. Before he got on to the jetty that fact had been communicated to some other person there, and in no time it was all over London, so that on the evening of August 1st, 1911, when I had a small dinner-party at my house, the country roads of Essex around us were thick with motor-cars bearing newspaper men wanting to know if it was true that I had found gold in Spitzbergen! I mention this to show the enterprise of those who are on the look-out for news.
>
> People naturally asked, "How is it that you have discovered it after all these years?", "Has nobody else ever been there?", "Oh yes," the answer was, "but they were Norwegians, and they only go to cut up whales and are frightened to death of going over the mountainsides for fear of seeing hobgoblins; they just go and cut up their whales and get back as soon as they can, they have no enterprise."

Large machinery still remains at Ny-London. Photo: Susan Barr

Jacobsen described, with hindsight it would seem, that it was not unusual to find occurrences of iron, copper, pyrite (iron sulphide), galena (lead ore) and other minerals but always too little and sporadic to be a basis for even the most simple mining project. In 1919 and 1920 he states that NEC engaged no less than three American and three English geologists with four Norwegian arctic ships, but even this effort could not conjure up any productive occurrences within the Company's 15 000 km² large occupation areas.

The 1912 expedition

The NEC sent a large expedition to Svalbard in 1912: 75 men, 40 of whom were English. Mansfield was again the leader, with Birger Jacobsen as assistant. The marble expert W. G. Renwick from London participated – as Hoel stated: "Known for his work *Marble and Marble Working*", as did marble expert R. Gibson who represented the Ingersoll-Rand Company of New York and John Kelly who represented The Sullivan Machinery Company of Illinois, USA. Both of these firms produced machines for marble production. This summer it was *Activ* of Haugesund that brought the expedition from London (22 June) to Kongsfjorden on 11 July. After being unloaded, *Activ* proceeded with the winterers from Blomstrandhalvøya to Bellsund where also the winterers from Camp Bell were fetched. According to the ship's captain, Th. Tangeraas, ten cases of gold were also loaded here. *Activ* then dropped the Norwegian winterers off in Norway and proceeded to Aberdeen where it loaded more materials for the marble workings (building materials, logs, machinery and coal). One would have thought, if Ernest actually did have interests on the south side of Kongsfjorden, that he would have utilised the coal occurrences there. These were considerable and they became the basis of the Norwegian Kings Bay Coal Co. that was established there (becoming the settlement of Ny-Ålesund) in 1916, and they lie straight across the fjord from Blomstrandhalvøya. The *Activ* also brought the NEC secretary J. R. Maples and the shareholders King Smith and Mortimer, as well as F. G. Gardner, described by Hoel as the expedition doctor. *Activ* sailed three more times that summer between Spitsbergen and Norway, calling at Bellsund each time (Hoel:436).

At Blomstrandhalvøya four larger and two small living houses were erected, so that there was now room for 70 men. In addition, a workshop, a storage building and a smithy were built and a five-ton crane was placed on a bank above the shore so that ships could be loaded direct from land. Another crane was placed by the quarry, and a 5 ft gauge rail track was laid between the two – jokingly called The Great Northern – to transport heavy machinery and marble blocks. Wells were dug and water pipes laid to the buildings, the quarry and the harbour. Some of the heavy machinery for quarrying the marble arrived late that summer owing to a dock strike in London, and there was thus only time to quarry and ship a small amount of marble back to England (Hoel:436–37).

COMPANY'S SHIP AT MARBLE PIER.

"Activ" by Marble Island. Source: Marble Island prospectus, courtesy of Richard Gardner

The cabins at Marble Island were given names after prominent men in the Company, here Maples and Lagercrantz. Source: David Booth, courtesy of Ross McNeill

Meanwhile, at Bellsund Arthur Mangham and Birger Jacobsen worked in the area prospecting for valuable ores and even gold. In addition they erected a large number of claim signs dated 1905 (Hoel's information p.438) on the claims around Bellsund and Hornsund. Ingvald Svendsen of Tromsø sold his claim area by Van Keulenfjorden to Mansfield 6 July 1912 – keeping the fishing and hunting rights in the fjord (around 1930, Svendsen moved an NEC cabin from Recherchefjorden to Van Keulenfjorden and established beluga catching in the fjord. Today a large pile of bones witnesses the fact that at least 550 of these small whales were slaughtered here in the 1930s) – and this area was added to the large area around Bellsund that Mansfield/NEC now claimed or owned (Hoel:438). There was a clash with another mining company in the Bellsund area this summer. The Norwegian geologist Gunnar Holmsen and Sam Eyde, founder of the company Norsk Hydro, established the company a/s Kulspids in 1909 and apparently claimed the asbestos

The channeler and crane at work. Source: Marble Island prospectus, courtesy of Richard Gardner

occurrence in Recherchefjorden. Birger Jacobsen and NEC shall then have unlawfully taken over the asbestos claim in 1912. a/s Kulspids continued, however, to assert their claim and also mined and sold a small amount in the next few years. In 1919 11 men employed by a/s Kulspids arrived at the site to find a group from NEC, led by Captain Frank Wild, Ernest Shackleton's Antarctic colleague, in full activity with 13 men, mostly Swedes. After a while the NEC men left the site and the Kulspids group could start their work there. There was a new incident in 1922 when Kulspids' caretaker at the site had a serious skirmish with a NEC group led by Carl S. Sæther from Tromsø. In 1927, after the Treaty of Spitzbergen had come into force, ownership of the claim was granted to a/s Kulspids. The NEC relinquished its claim to the area in connection with an agreement with the Norwegian state which, to a larger degree, concerned the Kings Bay/Kongsfjorden area (Barr 2003:70–71).

The southwest side of Ebeltofthamna today. Photo: Susan Barr

Winter at "Marble Island", Blomstrandhalvøya 1912–13

The 1912 expedition left Svalbard on 19 September, leaving a caretaker group at the marble works consisting of leader Herbert Millar and nine workers, including again the experienced trappers Henry Rudi, August Olafson and Johan Oxaas (Arthur Oxaas' brother). Also in the group of nine were Ernest's brothers-in-law David and James Booth, James acting as cook and David assisting with the core drilling. The Booth family came from Glasgow, Scotland. During the winter seven holes were drilled to various depths, and the marble cores that were obtained were polished and sent to London for analysis. Even Svalbard geologist Adolf Hoel had to admit that "they showed a good quality" (Hoel:439).

After working at the marble quarry for a while, Henry Rudi and August Olafson bought themselves a set of supplies cheaply from the mining camp and rowed to Kapp Mitra to hunt seals. Then, as the winter drew near, it was decided that Olafson should move to Mansfield's cabin Camp Zoe for the winter trapping season.

In 1911 a German meteorological station had been established on the Arctic Coal Company's property near Longyear City. After a year's experience it was decided to move the station to a more independent area, and Ebeltofthamna in

Cook at the stove – this could be James Booth.
Source. Ross McNeill

An old stove still lies in Ny-London today – but obviously not the same one that James Booth used.
Photo: Chris Wainwright 2011

Krossfjorden was chosen. The Deutsche Seefischerei-Verein's research ship *Poseidon* transported the scientific leaders Kurt Wegener and Max Robitzsch to the site, together with the two assistants Michaelis and Schwarz, and equipment and building materials. The purpose of the station was mainly to conduct aurora and meteorological observations. At the end of September 1912, a telegraph link was established to the Norwegian telegraph station at Finneset, Grønfjorden. This station was now the nearest neighbour to the Norwegian trappers.

As Olafson and Rudi rowed to Camp Zoe, they found that the Germans had erected annexation signs stating amongst other details: *In Namen des deutschen Kaisers*. Rudi told that they pulled down every sign they found. At Camp Zoe "there was even a sign right over that as well. That was when it burst for Olafson. He charged like an angry bull and shattered it to bits. Then we boiled coffee with the splinters" (quoted in Lønø 1998:109). A few days later they visited the Germans at Ebeltofthamna and told them what they had done. The Germans were not particularly pleased but said that it was none of their business, and the Norwegian and German winterers became good neighbours.

David Booth kept a diary (in private family possession) of his expedition experience from 19 September 1912, when the *Activ* left the marble camp, and to 23 July 1913, when the wintering party were relieved. This gives a more unusual insight from the point of view of one of the British participants. David starts immediately by noting that the marble was very brittle, but that they had hopes of it improving further down. After coring to 6 ft the marble still came up in small pieces, so Millar had them moved to another drilling site. A great many entries tell the same story – the marble was fine with various colours including dark red and blue, but even at 49 ft it still came up in small pieces. They had some trouble with water freezing to ice in the drill holes, but even in mid-winter David does not complain of the cold or dark. He does mention that a "great storm" on 29 January nearly blew the houses away and that the storm continued for several days. The camp was running out of "nearly everything", not least butter, and he wished for open water so that they could get hold of some seal meat. Rudi and Olafson went to the marble camp to celebrate Christmas and had an enjoyable time that became extended when wind broke up the ice in the fjord. On 5 February it was −35°C, but on the 15th there was obviously open water as there were "men out in boat trying to get some seagulls as we are in a very bad way for all sorts of meat. They got twenty two gulls which we will have for lunch tomorrow wish we could get some seals or bears". It can be noted above that the Norwegian trappers Rudi and Olafson had hunted in seal meat supplies earlier in the winter and probably managed well on that. By 20 February David is despairing of the marble: "Marble very bad, impossible to get good cores in stuff like this. 46 ft".

David Booth (2nd from left) operating a coring machine. Source: Ross McNeill

On 21 February an event occurred that was to give Mansfield a good deal of publicity in the aftermath: the Germans from Ebeltofthamna arrived to ask for help in rescuing the members of a German exploring expedition under the leadership of Herbert Schröder-Stranz. The rather badly planned summer expedition turned into a winter disaster where eight men died and seven others struggled for their lives under awful conditions. The ship *Herzog Ernst* got caught in the ice off Nordaustlandet, and Schröder-Stranz with three others left the ship taking a boat, kayaks, tents and other equipment to cross the ice to Nordaustlandet to carry out the planned exploration. They were not seen again. The ship moved westwards to lay out a depot but got caught in the ice for the winter on the northeast coast of Spitsbergen. There were six Germans and five Norwegians on board, it was the latter part of September 1912 and they were 300 km from the mining settlement Longyear City. Various attempts to walk to fetch help left some expedition members dead or badly injured and others spread over a large area. On 27 December the captain of the *Herzog Ernst*, Alfred Richter, staggered into Longyear City and the news of the tragedy was quickly telegraphed out to the world. Several search and rescue expeditions were gradually organised, but this was mid-winter in the high Arctic and it was not before 24 January that a dog-sledging expedition was able to cross the Isfjorden to head north. They were forced to return to Longyear City, arriving on 12 February with one man having frozen both legs and two sledges and several dogs lost. It was now that a telegram was sent to the German station at Ebeltofthamna. (For a detailed description of the Schröder-Stranz tragedy see Amundsen 1983).

The men at Blomstrandhalvøya were immediately eager to help and Mansfield had sent instructions by telegram to Millar – via Ebeltofthamna – that they should do all they could. Wegener, with Millar, James Booth and several others including Rudi and Olafson began by moving supplies over to the NEC cabin on Storholmen (= Large Islet. Named Davis Island by Mansfield) further in Kongsfjorden and thereby nearer to the frozen-in ship in Sorgfjorden. David Booth went to Ebeltofthamna to fetch provisions and also to help the doctor there receive all the incoming telegrams. Mansfield's men then struggled to lay depots up the glacier Kongsbreen at the end of the Kongsfjorden, since this was their only possible route to get towards northeast Spitsbergen. They had obviously not expected such an expedition and were not equipped for it, and the conditions were against them. After some days Wegener decided that most of the men should return and only he with Millar, Olafson and Abrahamsen should continue. They struggled as far as Gråhuken and could see across the large Wijdefjorden (the Wide Fjord) almost to Sorgfjorden but were unable to cross it owing to bad ice conditions. After further struggles and an impressive distance covered, they dragged themselves back to the marble camp at Blomstrandhalvøya. David Booth notes very soberly "30th [March] Sun. Mr Millar Dr Wegener and two men come back from Expedition today they did not find anyone in Wilde Bay [sic] so had to come back as their food was finished."

Mansfield could fill in these bare facts with considerably more dramatic detail, which he did in his unmistakable style in an article in the English magazine *Vanity Fair* on 20 August 1913. Entitled *A struggle with death in the Arctic – How a small colony tried to save the Germans*, it is from "Our special correspondent, Mr. Ernest Mansfield" and shows a photograph of Mansfield standing outside Camp Zoe and holding what the magazine said was a copy of *Vanity Fair* "(Taken in the light of the Midnight Sun.)". Mansfield starts by almost ridiculing the Norwegian relief expedition led by Captain Arve Staxrud, who had several years of experience in Svalbard, both as topographer and also leader of the annual Norwegian scientific expeditions (Barr 2003). According to Mansfield, it had started so late (mid April) that the spring was well on the way and therewith ample food from the birds and animals that were abundant at that time. The "gallant little band" from Port Peirson, on the other hand, was travelling in winter when the landscape was locked in ice and "covered with a seven-foot mantle of snow". The men had "Soft snow to

The NEC cabin on Storholmen in September 2009. Photo: Hilde Tokle Yri/Governor of Svalbard

travel upon and hail, rain, and snowstorms to travel through. Their ski acted like suckers and stuck fast in the snow. When they removed them they plunged up to their knees in the heavy going". On their return journey "Fatigue begins to tell its tale. Nature craves rest; but the cold compels them to move on. Some desire to sleep, but all know what it means, so they stave off death by keeping each other awake. They limp along. They are dejected, tired, famished. The days are forgotten, for everything now is the same. They are making for home, Port Peirson, Camp Davis [= Storholmen], a crevice in the ice, in death – anywhere". Mansfield ends the article by hoping that the Kaiser will recognise the efforts that Wegener, Olafson, Abrahamsen and Millar made. "Also it would not be out of place to give some mark of appreciation to James Booth and Johann Oxaas for their assistance in laying the depots, and to David Booth for the help he rendered at the German station in Cross Bay [Krossfjorden] whilst the leader was on the noble quest to help and relieve his fellow countrymen".

Kurt Wegener gave Mansfield's name to what is today called Mansfieldfjellet (Mansfield Mountain), northwest of Dicksonfjorden.

While the search expedition was away from the marble camp, David Booth recorded that the drilling activity continued, but still without success. He began to suspect that some of the workers were sabotaging the drilling by dropping a nail down the hole, which stopped the drilling machine, but he had no proof of this. At the same time he was getting worried about Millar and the others, since nearly

The Northernmost inhabitable Cabin, with Mr. Mansfield reading "Vanity Fair." (Taken in the light of the Midnight Sun.)

Ernest at Camp Zoe. Source: Vanity Fair of 1913, courtesy of Ross McNeill

a month had passed. On their return David accompanied Dr Wegener to the German base so he could fetch matches that the marble camp had run out of. In May, ice in the drilling hole was causing real trouble, and David, James and Millar took it in turns to get up at 05:30 each morning to light a fire to melt the ice so that drilling could restart some hours later. At 58 ft he thought the marble was "much more solid looking", and at the end of May they got an 8 ft core at 66 ft, just before the relief ship arrived on the 30th.

Work was not finished for the winterers yet. They unloaded 75 sacks of coal and distributed them to the various houses, shovelled snow from the railway track, helped to erect a new storehouse with railway track into it and fixed up a radio station that they hoped would manage to communicate with the German station in the next fjord. On 21 June new supplies arrived by ship including more coal and cattle. A total of 373 sacks of coal were brought to the marble camp the year the Booths were there. Several of the men were quite ill now at the end of the season, but they gradually recovered without it being clear what the cause was.

David and James Booth, Ernest's brothers-in-law, were neither mentioned nor thanked in the glowing reports that Ernest wrote after this season's work at the marble quarry. Perhaps David had been too critical of the marble, although the samples seemed to improve at the very end of the season. The brothers were, however, giving glowing mention in the *Vanity Fair* article as quoted above.

At the NEC camp by Recherchefjorden, trapper Gustav Lindquist continued as a caretaker for Mansfield and the NEC for the winter 1912–13. He was accompanied by Trygve Olsen. The same two wintered there the following year, 1913–14, but unfortunately Olsen died of scurvy on 4 March 1914 (Lønø 1998:107, 115).

The NEC in the newspapers and the Company Prospectus

A Norwegian newspaper cutting in the Norwegian Polar Institute archives (unfortunately the cutting does not show which newspaper this was) dated 1 August 1912 under the heading *Spitsbergen's riches – Marble, gold and precious stones?* contains the paragraph:

> Finally we mention Englishman Mr Mansfield's expedition. Several summers in a row he has been on Spitsbergen. His expeditions have been surrounded by secrecy, but the English press could announce last year or the one before that that he apparently had found gold and precious stones. This year is it said that he has discovered considerable marble occurrences and that he has travelled north with 30 workers and four prefabricated houses which are to be erected. We will no doubt soon hear news about him.

The main Norwegian newspaper *Aftenposten* 1 October 1912 stated that:

> Mr Mansfield, director of the Northern Exploration Company, arrived yesterday in Kristiania [Oslo] from Spitsbergen. In the Company's mines and marble quarry 70 men have worked this summer, and great advances have been made. Around the mines a whole small town has grown up. The work will be continued next year. Ten men have been left at the marble quarry. Mr Mansfield has during the entire summer hired the steamship "Activ" and he loudly praises the officers and crew. In addition he has used two motorboats. English and American engineers have overseen the establishment of the expensive machinery and a railway.

The same newspaper on 27 November 1912 took up the "Spitsbergen Questions" by professor Olaf Broch concerning the negative ways by which claims were made, for example by describing large areas on a claim sign. "The unreality of such a claim has already been shown in practice, since, despite this [a previous extensive claim] the English engineer Mansfield's coal mines are worked within the claimed area". In *Tidens Tegn* 26 August 1912 in an article entitled *Norway's coal mines – concerning the latest news from Spitsbergen – The real situation at the current mines* by Harald Hagerup we read that:

> It is, as is known, particularly on West Spitsbergen [now Spitsbergen] that one so far has found coal – The most important occurrences are found in Bellsund, Isfjorden and Kings Bay [Kongsfjorden]. In Bellsund Mr Mansfield has, as just mentioned, found precious stones and the red gold and in Kings Bay he has his marble works. Anker has coal occurrences here, which he means to start working in the nearest future. Mr Mansfield's name is well known on Spitsbergen. He shows initiative and stamina, but it is not easy to say whether his work will be crowned with the luck he probably deserves. People up there at least follow his gold digging with great scepticism. It is another question whether his marble quarry in Kings Bay will be more rewarding. It is said by those who understand these things, that the marble that might be taken out there most probably cannot be exported in large blocks. The severe frost will beforehand have shattered the stone.

It is perhaps strange that this latter prophecy, which turned out to be true, was not more widely spread at the time.

Tidens Tegn 19 July 1913 could report under the heading *To Spitsbergen after coal, marble and gold – An English expedition underway* that "The NEC, whose director is the well-known English prospector Dr Mannsfield [sic] from London, has equipped an expedition consisting of 46 Englishmen, of whom 8–10 engineers and the rest workmen …". The article is written in a hearty tone and tells how Blomstrand is just one block of marble and that the expedition will also prospect for gold. "The ship also has onboard a huge steam boiler, a steam engine, 72 tons of railway lines and 250 transport wagons. Everything indicates that the Company will considerably expand its production."

In 1911/12 the NEC published their prospectus *The Northern Exploration Company, Ltd*, with a magnificent photograph of Ernest posing as a romantic dreamer[2]. The publication contains a series of letters and reports that do not name Spitsbergen at all but hide the geographical location behind the code letter "N". The first statements – by Mansfield, Gardner and Salter – are addressed to Henry Williamson, the NEC's first chairman of the board. Ernest states in a brief memoir that he has already been seven years (summers) in "N", which dates the publication to 1912. Gardner states that they found gold in 1905 as well as plenty of coal

---

2   NEC published two books, one entitled *The Northern Exploration Company, Ltd* and the other *Marble Island*. They both appear to be shares prospectuses and are not dated but the letters within the former are dated 1911. It has the splendid photo of Mansfield at the beginning. *Marble Island* is also not dated but has a letter in it from Mansfield dated 1912. It also has the same photo in it, but of better quality print and it has many other photos as well.

and other valuable minerals. It sounds, in fact, as though they had discovered an Aladdin's cave in this secret place: "there is enough coal there to supply Europe", he claims, adding that "(M) is straight beyond question". Salter agrees that having known Mansfield for eight years he has "found him straight as a die and I believe him thoroughly in all he says". He also gives the impression that Mansfield has "gold fever" and that he wintered in the area in order to find gold, whereas Salter himself had more belief in coal than gold. Ernest reports on the marble occurrences which he states that he discovered in 1906, calling the area Marble Island. "During my investigations of this wonderful Island, I was so impressed with the value of the discovery, that I kept the find a profound secret from all the foreigners in my service, and from the ship's crew". George Alexander states that in 1906 they had their ship *Mulygan* on the other side of the Bay (fjord) and that they took a boat across to Marble Island. In addition, Charles Mann mentions that the group was on the south side of the fjord (where Ny-Ålesund is today) and that they washed a little there and found traces of gold, although it seems inconceivable that Ernest would have passed up occupying an area with traces of gold. The publication is altogether full of diverse statements about the marble occurrences. Olaf Martinsen of Tromsø, Ernest's interpreter in 1905 and 1906, states that "I would go with Mr Mansfield anytime, he has always been spoken well of by all his men. It is greatest pleasure for me to say this". For the coal occurrences at Kolfjellet Mansfield stated "The value of this coal area must be reckoned in millions sterling". Ernest mentions also that oil may be found there. Photographs on pp. 124–28 show mine shafts with inscriptions such as "This is a small shaft I sunk south side of Kings Bay with 16 men. Staked off 1906 and worked by me and 16 men, 2 English & 14 Norwegian".

Regarding gold Mansfield states in the Prospectus: "Although I proved the existence of these great gold-quartz reefs so long ago, my whole time since has been occupied in opening up coal measures in other places …". He had not found the opportunity to continue with these finds, but had visited them each year to make sure that no one else had found the place. He stated that he had been in Alaska, New Zealand, Africa, Canada, America and Australasia. Although it is unclear in the Prospectus to which area/country Mansfield is referring regarding the gold reefs, this is obviously the north side of Bellsund. He wrote about these Banket Gold Reefs: "In 1908 I devoted the whole of my time to prospecting for the Reefs yielding this stone, and in October found them in a place where few prospectors would waste time in looking for them. I was so impressed with the discovery, that I refused to leave the field, and remained for a year and ten days on the spot where the discovery was made. In the following June I returned to England. At the present moment on one camp I have eleven men working on the Reefs". It was in 1908–09 that he wintered at Camp Bell. Further, "For seven years I have been persevering in my quest for the rarer minerals. I have unearthed the parent reefs that yielded the fragment Banket specimens carrying gold in paying quantities".

Svalbard geologist Audun Hjelle, with many years experience in Svalbard, states that it was by Millarodden (Millar Point) and Reiniusholmene (Reinius Islets just off the north coast of Bellsund) that Mansfield thought he had found traces of gold equal to the best in South Africa. On one of the islets, Sinkholmen (Zinc Islet) zinc is found in the form of sphalerite and a little galena and some copper minerals are found with this (Hjelle 1993:74). These islets were most probably the "reef" referred to by Mansfield where he thought he saw glimmers of gold.

A red book entitled *N.E.C. Private and Confidential* and apparently published in 1913 contains a listing of assets and goals, etc for the Company. W. G. Renwick's report after two months on Blomstrandhalvøya tells of large amounts of fine marble as well as the buildings and activities. Nairn's report of the marble samples he has investigated states that "Taking into consideration that your marbles are all from top layers, they are remarkably sound. Some of the Breccias are the soundest I have ever seen and should improve when lower layers are worked", a statement which seems to contradict what is known of the fragile nature of the marble. Further that "Many of your marbles … from their beauty they would create a market for themselves". R. Oswald Lamigeon "… I would mention that I consider the particular merits of your Marbles, … are their soundness …". There were further words of praise about the marble after investigation by experts dated 1912 and 1913. However, someone (probably Adolf Hoel who led the Polar Institute at that time – the copy of this prospectus in the Norwegian Polar Institute library has Birger Jacobsen's signature across the inner front cover) – has placed question marks in the margin against some of the statements. For example: "In addition traces of oil have been found in Lowe Sound" is heavily underscored and supplemented with two exclamation marks.

There are reminiscences of the praise in a report from Ernest's prospecting activities in British Columbia (see the chapter *Ernest in Canada*). In the *Nelson Daily Miner* of 18 September 1900 one can read that "The Ernest Mansfield Manufacturing Co. has prepared a beautiful piece of marble and building stone to be sent to the Spokane exhibition. The sample is a good one and undoubtedly the best marble and work that can be had in Eastern British Columbia".

Another book (grey this time) with the same title *N.E.C. Private and Confidential* (probably 1914) starts with Birger Jacobsen's words of praise about the iron occurrences. After four months work in the field there, he can state that Martinfjella – on the east side of Recherchefjorden – is a solid range of ore of good quality. Recherche Bay can quite easily be kept open all winter, he adds, which is an amazing statement from one who should have been familiar with the high Arctic conditions in Svalbard: he was there for the first time in 1909, before working for the NEC. Further that there were great visions of easily earned mineral riches in the area. Many pages of equal praise concerning the iron occurrences continue, followed by praise of the coal occurrences and of the marble. The book finishes with colourful photographs of various marble samples. It would seem that Birger Jacobsen at least gave very conflicting views about Mansfield and the NEC according to whom he was speaking to or on behalf of at the time.

Marble Island and its potential
The NEC had invested a considerable amount of money in the marble quarry and the site was fully equipped with machinery and buildings. The quality of the marble specimens that had been brought to England had been praised to the skies by experts from Britain, France and USA, and the experts who had been to the site in Svalbard had added to the praise. The quality and quantity at Marble Island and neighbouring areas were excellent and of a great variety. Marble expert Mr Renwick wrote in his report that he was well acquainted with many types of marble from many parts of the world, but had nowhere "met with such variety, nor have I seen or heard of a Marble proposition possessed of such exceptional facilities, both for quarrying, handling and distribution. I am quite convinced that in the Marble formation of Kings Bay, the Company possesses by far the most valuable deposits of Coloured Marbles existing. ... Marble so beautiful, so varied, so highly decorative, so adaptable to every purpose where marble can be used, must become of World-wide demand". The value of the property "is incomputable" (report of 17 March 1913, quoted in Hoel:440). John Kelly, who had also spent several summer weeks at Blomstrandhalvøya in 1912, was of the same opinion (report of 30 September 1912 to NEC, quoted in Hoel:441). David McEwan, foreman for the quarries in 1912, wrote to the NEC director 5 August 1912 that "In a number of places the frost and atmosphere have helped to damage the surface stone", but this was – he believed – only a surface problem (quoted in Hoel:441). Hoel admitted that there were many extremely positive analyses of the marble at Kongsfjorden, also from a number of other marble experts in Britain, France and USA who had been sent samples and had cut and polished them. Not least the large variety of colours and character amazed these experts. The praising reports were published by the NEC in April 1913 in a book entitled *Marble Island* with 28 testimonials, 80 photos of their work on the Island and 30 colour photos of marble.

Adolf Hoel actually admitted that all the positive reports "made it understandable that the Company invested so much in the marble at Blomstrandhalvøya". In

Marble sample colour plates from the NEC Prospectus, courtesy of Richard Gardner

addition "There was no lack of strong, flattering statements from experts about the quality of the marble, the quantity and the competitive advantages. The Company's leadership therefore had good reason to maintain that rational working of the marble field would produce a satisfactory result". Supportive words indeed from Hoel.

The 1913 season

On 31 March 1913 Johan Hagerup sold to Mansfield and the NEC for NOK 1500 all the ten hunting cabins he had erected in Svalbard during the years 1897–1910. One was said to be on Tusenøyane although there is no indication of a hut in that area, two at Hornsund, and the remainder in the Bellsund area. This can be interpreted as an indication of faith in or the esteem in which he held Mansfield after many years of cooperation. On 27 May 1913, Charles Edward Evans joined the Board, while on 1 October the same year Warburg resigned.

Ernest led a new expedition to Svalbard in summer 1913, accompanied by Birger Jacobsen and the famous English polar expeditioner Cmdr Victor Campbell, who had led the Northern Party on Captain Scott's fatal South Pole expedition 1910–13, and took over the command of the whole expedition when Scott was found to have perished. The trip to Svalbard was in fact remarkably soon after his return from Antarctica. Again the steamship *Activ* was used, as well as the two motor cutters *Cynthia* and *Zoe*. *Activ* made seven trips between Tromsø and Svalbard that summer, each time calling in at Hornsund and Bellsund. A new large storage building was erected at the marble workings and more rail track was laid from a new marble quarry. Hoel remarks that it did not seem that much marble was actually quarried this summer either (Hoel:443). Jacobsen and Campbell used the *Cynthia* – skippered by Johan Hagerup – to visit NEC's various properties from Hornsund to Bellsund. A cabin was erected by Van Keulenfjorden and another at Camp Morton next to the Michelsen building. Jacobsen also claimed for the NEC an area of mid-Spitsbergen east of Prins Karls Forland, from St Jonsfjorden to Trygghamna at the mouth of Isfjorden, and erected a large number of claim signs (Hoel:443–44). At the end of the summer all the workers were sent back home with the exception of a small group who were left to look after the workings through the winter.

The NEC decided at an extraordinary general meeting 25 July 1913 to raise the share capital from £125 000 to £150 000 by issuing 25 000 new shares at £1.

Something happened this summer that seemed to have had a very negative consequence for Ernest. American mining engineer and marble expert Mr Minard, together with assistants Sherer and Barnett, arrived at Blomstrandhalvøya to make a detailed assessment of the situation there. As a result of Minard's report, Ernest was removed from all involvement with the NEC. Hoel remarks (Hoel:444) that it was not necessarily the marble workings that were the cause since the NEC did not give the site up. He speculates that it was Mansfield's leadership that was the problem. He quotes the obituary that Ernest's employee of several years, Birger Jacobsen, wrote in the Norwegian newspaper *Tidens Tegn* after Ernest's death in December 1924 where he said amongst other things:

> Mansfield was of a magnificent character with a strongly developed poetic vein and an exaggerated imagination which is often visible in men who have made the search for useful minerals [Jacobsen wrote only "exploration", not "the search for useful minerals"] their life's goal. ... He has with honour written his name into Spitsbergen's history, and he will always be remembered as one of the archipelago's first and most important pioneers (Hoel:444).

In a legal statement made in Tromsø in 1922 Mansfield himself stated that he had no more connection with the NEC after that summer. However, he was clearly not removed entirely and was heavily involved in the *Marble Island* book produced in 1913 and, although neither he nor the Rev. Gardner are listed in it as directors, their photographs adorn the introduction. It just seems that he did not go back to Svalbard. It may be speculated that he had made so much money that he did not need to. He was 51 years old in 1913 and only lived to the age of 62, so perhaps he was not fit enough to go. Smoking, coal dust, frost bite, etc. may have damaged his health. Not least, World War I broke out in 1914 and it was not possible to go to Svalbard during those years. Later statements in the newspapers that he addressed general meetings of the NEC "at some length" (*The Times* 17.12.1919 and 7.4.1920) show that he continued giving presentations about the NEC and its assets and prospects.

The first expedition without Ernest (1914)

With Ernest now out of the picture as regards expeditions, the one in 1914 was led by Victor Campbell, with Birger Jacobsen, and the Company had now bought its own ship, the *Willem Barents*. Johan Hagerup was the ice pilot. The claims at Bellsund and Hornsund were visited without any work being done. However, the area to the east of Prins Karls Forland, that Birger Jacobsen had claimed for the NEC the previous year, was examined for coal and other minerals. The news of the outbreak of World War I in August gave an abrupt end to the expedition. The properties at Blomstrandhalvøya and Bellsund were inspected and catalogued and, although the expedition had been ordered to proceed direct to England, it was decided it would be safer to go first to Norway, where they arrived 21 August (Hoel:444–45).

Victor Campbell then devoted himself to the war effort, during which he distinguished himself several times.

Ny-London on Blomstrandhalvøya today. Photo: Susan Barr

The World War I years

No work was done at Kongsfjorden either this summer, but two men were left there as caretakers for the winter 1914–15. Again one of these was the famous polar bear hunter Henry Rudi, the other Arthur Oxaas' brother Johan. Arthur Oxaas wrote in his book (Oxaas 1955:101) that they had been promised supplies, but these never arrived. Rudi became impatient and decided to travel back to Tromsø with a post ship to find out what had happened. In fact, World War I had broken out. The men were left with no contract and no equipment, and the Englishmen packed and left. Rudi went with them to Tromsø to sort out matters, but there was no contract to be had this time. So he stayed in Tromsø. Arthur Oxaas was in Grønfjorden when he heard about this state of affairs, and he decided to winter with his brother Johan. The Germans in Ebeltofthamna were also evacuating, and they asked Oxaas to look after their station since he would anyway be wintering in the area. The two brothers decided to stay in Ebeltofthamna and to use the marble camp as a secondary station. Oxaas remarked that the German station was of very high standard. They left for home towards the end of August 1915 (Lønø 1998:119).

For the winter 1916–17 the NEC hired the trappers Mikael Olsen and Mattson as caretakers at Camp Bell. The following winter 1917–18 the NEC contracted Olsen, Mattson and Ingvald Nilsen to look after the coal workings in the Bellsund area. Most of the NEC huts and cabins in the area were used either as main stations or as secondary stations for the trapping activities during the winter. The Swedish mining camp at Sveagruva helped with doctoring of both Mattson and Nilsen during the winter. They got passage home with a fishing boat 19 June, but Mattson unfortunately died onboard before they reached Norway (Lønø 1998:123–24).

During the war, there was no possibility for further NEC expeditions to or activities in Svalbard. Birger Jacobsen chose to interpret this as a breach of contract and that he therefore was released from obligations to the company. In autumn 1915, he therefore transferred his occupations to a Dutch company and accompanied their expedition to Recherchefjorden in summer 1915 where the iron occurrences were inspected. Expert opinions given between 1912 and 1916 about this occurrence declared it to be without economic potential, and the Dutch company did no further work there (Hoel:445–46). At a meeting in the Royal Geographical Society in 1915 (reported in the *Geographical Journal* 1915), however, Antarctic and Svalbard expert Robert Neal Rudmose-Brown, also a consultant to the mining company Scottish Spitsbergen Syndicate, gave great praise to the marble occurrences at Blomstrand and iron at Recherchefjorden.

The post-war years

The Peace Treaty of Brest-Litovsk that was signed 3 March 1918 between the German and Russian governments also contained the clause that the Soviet Union would support the German plan for the future governance of Svalbard. This, together with rumours that NEC claims had been taken over by foreigners, worried the NEC, and the Board decided to send an expedition to the archipelago to secure their claims even while the war was still underway. The plans were discussed with the Foreign Office and the Admiralty. In early spring 1918, Frederick William Salisbury-Jones was appointed managing director of the NEC with a salary of £6682 (A.G.E. Jones), and he was able to tie other influential names to the company. Salisbury-Jones' marketing of the company brought in new capital through the sale of 350 000 new shares à £1. The capital was thus raised to £500 000. Sir Ernest Shackleton and his Antarctic second-in-command Frank Wild were recruited to lead the expedition. Wild was released from the Royal Navy Voluntary Reserve on 5 July 1918 (A.G.E. Jones). Amongst the other participants were Salisbury-Jones, Norman Carlyle Craig KC, MP, Noel B. Davis, son of shareholder Fred. Lewis Davis, and Arthur and Bert Mangham, who were already experienced in Svalbard mining (see above).

According to Hoel (1966:447–48) the approval of the Foreign Office and the Admiralty for the importance of the expedition manifested itself in the loan of the armed vessel *Ella*, which left London at the beginning of August 1918 and was escorted in convoy over the North Sea. In Tromsø the ship had to wait for Salisbury-Jones and Craig, who had travelled separately to Kristiania (Oslo) to discuss the situation with the British minister and the Norwegian Foreign Ministry. On their way north to Tromsø they were met by Shackleton, who informed them that he had been called back to England by the government in order to take command of a British military force in north Russia. Captain Wild was to take over leadership of the Svalbard expedition. According to A.G.E. Jones, it was Salisbury-Jones who took over control, while Wild was to be manager in Spitsbergen (his information is from *The Times* and *Outspan*). Arthur Mangham also had to back out, owing to an accident causing him a broken rib. Herbert Ponting, famous for his photographs of Captain Scott's last Antarctic expedition 1910–13, accompanied the expedition. Before going to the Antarctic, he had been a mineworker in the USA (Wikipedia). A.G.E. Jones states (from *Outspan*) that included on the expedition were Dr J. A. McIlroy – who had served on Shackleton's *Endurance*, three Yorkshire mining foremen, a cook-steward, storekeeper and 60 Norwegian and Swedish workers. The voyage was not entirely without drama, since the war was still on and the Germans were still sinking ships.

The Times of 27 June 1919 recounts a different story, by referring to the speech held by the chairman Frederick Lewis Davis at the NEC Ordinary General Meeting. He ran through the history of the NEC, stating that when the War broke out in 1914 he spoke to Chancellor of the Exchequer David Lloyd George about the enormous deposits of iron ore in Svalbard and the importance of this for the country during the time of war and received support from Lloyd Davis. After some months' delay, the Ministry of Munitions then arranged for a ship to take engineers to Svalbard but, Lewis Davis recounted, he was greatly surprised when the Treasury refused to grant the money. Lewis Davis stated that it was Salisbury-Jones' ability to raise capital in 1918 that led to the expedition taking place that summer. The escort that the Admiralty had promised for the *Ella* never arrived in time and indeed the *Ella* had reached Svalbard before the escort arrived in Tromsø. Lewis Jones then briefly mentioned the fact that the German wireless station at Ebeltofthamna was dismantled by the NEC expedition and the property claimed for Britain (see below). It may be that Lewis Davis was exaggerating the NEC's role in the matter.

The *Ella* was a relatively fast ship and was fitted with an up-to-date 4 inch gun to repel submarine attack. She had 60 rifles and 100 000 rounds of ammunition in case there should be trouble with the German settlement in Spitsbergen, which in fact proved to have been abandoned (A.G.E Jones).

During a discussion about "The Political Status of Spitsbergen" in the Royal Geographical Society in February 1919, Herbert Ponting showed a series of photographs taken during the 1918 expedition and added that:

> The Northern Exploration Company, which sent out the expedition that I accompanied this summer, owns approximately 2000 square miles of land in Western Spitsbergen, a large portion of which territory is exceedingly rich in coal and iron. The country far exceeded my most sanguine expectations. I have seen great tracts of land where the coal-seams can easily be traced by the eye…and at Recherche Bay I tramped along the side of a mountain several miles long and some 1500 feet high, of which a large part seemed to be a solid mass of iron ore … of their [Spitsbergen marbles] surpassing beauty there can never be any question. Copper pyrites, molybdenite, gypsum, galena, and asbestos are known to exist in large quantities. (*The Geographical Journal*, Vol. 53, No. 2 (Feb. 1919), pp. 91–96.).

Ponting had much more to say, extolling the amazing natural advantages of Spitsbergen, both as a mining paradise and also as a future summer health resort "where plans for a hotel at Recherche Bay were already being discussed." A.G.E. Jones pointed out the fact that Ponting was a major shareholder in the NEC, holding £7000 in shares, "which he wisely reduced to £1050 as the fortunes of the company fell". Sir Aubrey Strahan, Director of the Geological Survey, joined the discussion to lay a damper on the enthusiastic reports of iron ore occurrences. He stated that no earlier geologists, including Swedish and Norwegian experts, had reckoned the iron ore by Recherche Bay to be worthwhile mining, both with regard to quality and considering the difficult natural conditions of this high arctic site. Chairman of the NEC F. L. Davis commented on this in *The Geographical Journal* of March 1919 (p.207), where he stated that "the analyses of the iron ores brought home by the officials of this Company in the autumn 1918 gave on assay the following results: [between 59.04% iron and 68.2%] thereby fully confirming the previous analyses reported by our engineers in 1913 and 1914".

A correspondent from the London newspaper *The Financier* also accompanied the expedition and wrote a series of articles that were published between 8 October and 12 November 1918. These were collected under the title *Spitsbergen's Mineral Wealth*, which comprises a summary of the 11 articles. The expedition members inspected the iron claim in Recherchefjorden and declared it to transcend "their most optimistic expectations of the extent and value of the iron mountain deposits". This is the occurrence which Adolf Hoel could explain had already been written off by several other experts (see above).

In Kongsfjorden the NEC expedition discovered "Norwegian invaders on land claimed by the Northern Exploration Company. Four ships lay in the roadstead – three steamers and a sailing ship; and on the shore two fussy locomotives pulled loads along a short railway track" (*Spitsbergen's Mineral Wealth*). This was the mining site opened by the Kings Bay Kul Comp. in 1916 and then named Ny-Ålesund. The NEC expedition was obliged to buy coal from the "Norwegian invaders", before they moved on to the German meteorological station that had been established at Ebeltofthamna in nearby Krossfjorden in 1910 and evacuated in 1914. There they pulled down the German occupation posts and put up their own, including hoisting the Union Jack. The Blomstrandhalvøya site was also inspected and proved to have been ransacked and partly destroyed. Nathan Haskell Dole in his book about the Arctic Coal Company in Longyear City also recorded that the marble works at Blomstrand were plundered during the war (Dole 1922:450). Kings Bay Kul Comp. agreed to pay £750 for goods they had taken.

Back in Recherchefjorden several tons of iron ore were collected and hopes were high for the value of the occurrence. Even Hoel gives the NEC the benefit of the doubt here and states that they must have been genuine in their belief that this in fact was a viable and lucrative claim. He writes that they were badly advised by experts who obviously did not have sufficient overview or competence to judge the occurrence. Indeed, Arthur Mangham stated on the background of his many years' activity in Svalbard that "My unshakable conviction is that the iron deposits at Récherché Bay are unexampled in the world, and practically inexhaustible. I said years ago that they were worth the Bank of England, and I say so still" (Hoel:451). Those on the 1918 expedition confirmed this opinion and their report stated that "Some of the stones, judged by its weight and appearance, were apparently pure iron". Hoel reckoned that NEC's people obviously were not acquainted with the report of the Norwegian mining engineer Mørch-Olsen, together with Birger Jacobsen and Nils Brandberg, which was far more negative about "Iron Mountain" (Hoel p.450–51).

Finally, the expedition moved over to Camp Morton, where the main winter camp was to be. Prefabricated houses, tools and provisions were unloaded for 50–60 winterers and work to prepare the camp began. Although the iron occurrence by Recherchefjorden was thought to be the most lucrative, it needed the establishment of a cableway for moving the ore to the shore before the real mining work could begin. The coal occurrences on the north side of Van Mijenfjorden by Camp Morton, could, on the other hand, be worked immediately. It was to the Swedish mine in Braganzavågen they had to go, however, for bunker coal for the *Ella* to return to Norway, and here the NEC people were impressed by the construction work and organisation of the mining site.

About 50–60 men were left to winter at Camp Morton – renamed Davis City – under the leadership of Captain Wild. It was stated that "What Juneau is to Alaska,

Davis City will be to Spitsbergen, and, with its superior advantages of accessibility and climate, it should become an important factor in the world's industrial history" (*Spitsbergen's Mineral Wealth*). The *Ella* joined a convoy at Bergen and returned safely to England (Hoel:446–453). Wild stated many years later (recounted by A.G.E.Jones from the *Outspan*) that "the standard army huts they had brought with them were totally inadequate and it took a long time to make them moderately weatherproof, even when they were covered with snow". The coal was carried down the hillside by a "flying fox" or zip-line to a light railway track for transport to a dump on the shore. Apparently, about 5000 tons were worked out (*Outspan*). Jones recounts further from Wild that he discovered that the summer shipping season was in fact only about three months long. However, once the fjord froze, it was a good means of transport to the Swedish mine at Sveagruvan 39 miles away. The winter weather was harsh, but the camp had a fairly extensive library and Wild continued the Antarctic tradition of concerts, usually on Saturday evenings. Hunting for foxes, ptarmigan and grey geese was also an option when the spring returned, and a polar bear was shot by Wild and McIlroy. Jones notes that "The health of the party was good except for the usual minor industrial injuries. One man who was sick when he embarked died during the winter and one man became insane" (originally from *Outspan*).

Meanwhile, back in England, the NEC was busy promoting its assets in Svalbard, stating that all the claims had been inspected by British and Norwegian experts "all of whom declared that the iron deposits are incomparably rich and unsurpassed in Europe" (*The Times* 1 October 1918). They maintained that over 400 miners had worked there during the 1917–18 season and there were twice that number in 1918. This at least was an outright exaggeration. The output in 1919 was expected to reach 100 000 tons. In spring 1919 an illustrated pamphlet was issued (*The Times* 16 April 1919). For the Annual General Meeting of the Company in 1919, *The Times* published (27 June 1919) Chairman Frederick Lewis Davis' statement over a full page and the activity seemed to be flourishing. Frank Wild recalled, however, (quoted by Jones from the *Outspan*) that "The expedition was not by any means a financial success". Salisbury-Jones replaced Wild with Army officer Ltn.col. A. D. Borton, VC, CMG, DSO, who left General Ironside in Archangelsk to become local director in Svalbard (*The Times* 27 June 1919). Wild was not unhappy to leave the NEC, but he made the mistake of not selling his £500 worth of shares. He still held them when the Company was dissolved in 1929 (Jones).

In his speech to the NEC AGM, Lewis Davis also played down the natural challenges of winter darkness and shipping problems owing to long periods of ice cover. In his opinion, the ice season was getting shorter and travel to Svalbard was now no more difficult than sailing from Hull to Kristiania. In addition, there was a new method of transport being developed that would completely solve the ice problem; that is airplane or dirigible. Even today, he stated, it was possible for a machine to fly from Tromsø to Bellsund in three to four hours (*The Times* 27 June 1919). Interestingly enough, in the same speech Lewis Davis stated that Ernest Mansfield was one of the leaders of the prospectors on the 1919 summer expedition. Local Director and Administrator of the properties in Svalbard Ltn.Col. Borton had already left for the archipelago.

The NEC board was now composed of the following persons:
Sir Harry E. Brittain, London
Frederick Lewis Davis, Surrey
Charles Edward Evans, Somerset
H. L. F. Lagercrantz, Stockholm (as a result of an agreement with the Swedish company AB Isfjorden-Belsund that operated in the Braganzavågen area)
Sidney Thomas Peirson, Coventry
F. W. Salisbury-Jones, London
Gerald Dudley Smith, Worcester

The time had come when the question of assigning sovereignty over the archipelago to one particular nation was firmly on the table. The World War was over and Germany was not a contender. This was a question for the winning powers to debate and try to solve. Nevertheless, in February 1919 the board of the NEC took the initiative for a conference of four of the companies that were involved in Spitsbergen. They invited representatives for the Norwegian company Store Norske Spitsbergen Kulkompani a/s (which had taken over Longyear City, now Longyearbyen), the Scottish Spitsbergen Syndicate and the Swedish Aktiebolaget Spetsbergen Svenska Kolfält. The invitation was refused, not least because Store Norske maintained that firstly a conference about mining conditions should take place in Kristiania and not in England, and secondly that the larger issues were a matter for state parties, not private companies (Hoel:453–54). The invitation does show, however, that the NEC felt that they were in some way a privileged part in the discussions about the future administration of Svalbard.

The 1919 expedition
For the expedition in 1919, the company was able to use the *Sheila* of London, owned by the first Earl of the North Sea and of Brooksby, Admiral of the Fleet, David Beatty. In addition the company used the *Kristoffer Ellingsen* of Sigerfjord and two motorised cutters. Several geologists participated, as well as Birger Jacob-

sen, who had decided to rejoin the company. Photographer Richard N. Speaight also participated. He was entitled "royal photographer" and was perhaps employed or invited on this expedition by the NEC for promotion purposes. One group was put ashore in Recherchefjorden to make a geological investigation and appraisal of the company claims in the Bellsund area, and a photograph by Speaight which was published in *The Graphic*, November 29, 1919 showed Salisbury-Jones investigating asbestos occurrences by Recherchefjorden. Others of the expedition group were transported to Blomstrandhalvøya and the marble workings. They found the site even more plundered than in the previous year, since no caretakers had wintered there. The only possible activity involved making an inventory of the equipment, since not enough was intact to allow any mining of the marble to take place. New equipment and spare parts did not arrive before 7 September with the *Kristoffer Ellingsen*, so that work to mine the marble could not start before the winter. Only a small force was left there for the winter 1919–20 (Hoel:453–54).

Also participating on the expedition summer 1919 were geologist Charles Boise with assistants who wrote a report about the economic possibilities of the various NEC claims. According to Hoel (p.454), they established that the ore occurrences at Martinfjella (Martin Mountains) on the east of Recherchefjorden were without any economical prospects, nor were the lead, zinc, copper or pyrites worth extracting commercially. The asbestos and graphite were a possibility, but the coal in the area, including Kolfjellet (Coal Mountain) was not worth mining. The marble at Blomstrandhalvøya was, however, promising; there were large amounts and many attractive varieties, the mining conditions seemed favourable and it was worth taking out new samples. There were also certain other areas with mineral occurrences that might hold some promise. In light of the very enthusiastic reports from summer 1918, the new analysis in 1919 was – although not hopeless – in fact rather a let-down, particularly with regard to the prospects of amazing iron ore occurrences in the Recherchefjorden area (Hoel:455–56).

On 27th June 1919 *The Times* of London printed a very long, perhaps full page report of the Northern Exploration Company's General Meeting (Times 27.6.1919). The report included the following information:

> Extensive mineral deposits found, including "the enormous and beautiful marble found at Kings Bay".
>
> Discussions with Prime Minister Lloyd George and the Ministry of Munitions about ownership of the land.
>
> "Mr Mansfield has spent many winters on Spitzbergen and threw himself with characteristic energy into the scheme, and advanced considerable sums of money…to procure a ship and fully equip it with food and machinery that at times looked like an almost impossible proposition".

The 1918 expedition under the command of Sir Ernest Shackleton… "One of the results of the 1918 expedition was the extension of our properties by taking German territories, the dismantling of their wireless station, and hoisting the Union Jack on their flagstaff".

Soon after Mansfield had discovered the "Iron Ore Mountain" by Recherchefjorden a Dr. Voight of Berlin had promised to make him a fortune if he sold the property. Mansfield had refused, stating that he was a "Britisher" and anything he discovered in Spitsbergen was for his own country.

A claim that the "beautiful marble" was worth £4 to £16 per ton. "The supplies appear to be inexhaustible and the varieties wonderfully varied in colour".

Trespassing was "rather flagrant" during the war. The British government made strong protests to the Norwegian government.

The Company directors had entered into a contract with the Marconi Company for the installation of wireless stations at the properties at Kongsfjorden, Recherchefjorden and Lowe Sound (Van Mijenfjorden), and that these were nearing completion. [These were in fact either never installed, or were never used successfully].

After the 1918 expedition the company owned 2000 square miles of land, since then it has considerably increased.

The 1918 expedition provided huts and equipment for 100 men. The 1919 expedition will "provide accommodation and equipment for at least 500 men".

Lt Colonel Borton VC has been appointed Local Director on Spitzbergen on the recommendation of the War Office.

The article also included a map of Spitsbergen showing the area of land claimed by the company.

The historically well-known Svalbard trapper Gustav Adolf Lindquist, who had been wintering caretaker for Mansfield in 1911–13 and had also assisted the NEC expedition in 1919, had apparently occupied in July 1918 an area on the southeast of Spitsbergen where he claimed there was an interesting coal occurrence. He called the natural harbour there Davis Harbour after Frederick Lewis Davis, board member of the NEC. This site (Davishamna in Norwegian) has now silted up and is no longer a "harbour" (Place Names). On 17 June 1919 he signed a declaration in Tromsø which stated that he had claimed this area the previous year and that he now sold the entire occupation to Ernest Mansfield for the sum of 1000 kroner in cash. This was confirmed by Mansfield on the same document and witnessed by two persons in Tromsø, the one being Carl S. Sæther, NEC's representative in Norway. Adolf Hoel is the detective on the trail here. He expressed disbelief that Mansfield could have been in Tromsø in 1919 to sign a document and have it witnessed. Mansfield had himself declared on 25 May 1922 that he had not led an expedition to Svalbard for the NEC since 1913 and that he had been removed from all connection with the NEC that autumn. Hoel could not see that Mansfield after this had been involved in NEC affairs, although we have seen that he in fact was. Hoel reasoned that the contract had been sent to London for Mansfield to sign there. In a contract of 8 July 1919, Mansfield then transferred Lindquist's claim on to the NEC, who used the documents to support their claim to the area (Hoel 457–89).

Also this summer Birger Jacobsen claimed the rights to gas deposits by Grønfjorden, which had been discovered the previous year by an engineer at the Finneset telegraph station. Jacobsen erected two plaques in the name of the Arctic Oil Co., which in fact was a daughter company under the NEC. In 1920 the NEC set up a drilling rig there, but Store Norske Spitsbergen Coal Company personnel appeared the day drilling began and protested against the trespassing on their claim land. The NEC men were told they were heavily outnumbered, and they ceased drilling after verbal protests; the rig was dismantled and loaded on to a boat (Hoel 1966:456, 463–64).

The 1920 expedition

Three of the board members – Davis, Evans and Lagercrantz – withdrew at this time and were replaced by Harry Stewart, Arthur Barton, the Hon. Edmund Colgerhoun Pery and Director Alf Frantzen of Kristiania (Oslo). Optimism was still the word despite the rather dampening geological report from Charles Boise, and the company capital was increased 100% with the issue of 500 000 new shares at £1 a share. The expedition in 1920 used five ships and sent separate groups to the various claims to investigate the diverse mineral occurrences. During the winter some coal mining had taken place at Calypsostranda by Recherchefjorden. Five men were landed by "Copper Camp" at St Jonsfjorden to start taking out copper, which had been discovered there the previous year. Attempts were then made to get to Davis Harbour, Van Mijenfjorden and Kongsfjorden, but all were stopped by ice. At the end of May – which by normal standards was very early for shipping activities in this area – they put men ashore at Peirsonhamna, Blomstrandhalvøya, under the leadership of mining engineer Kenneth L. Gilson, who had previously worked for the American Arctic Coal Co., now sold to Store Norske Spitsbergen Kulkompani a/s. Birger Jacobsen and engineer Bevan were put ashore at Camp Zoe to prospect there.

Later in June the main expedition base was established at Calypsostranda with Gilson as leader. Others in the group were G. Lash (wireless operator), H. G. Elston (assayer), G. Lowdnes (storekeeper), F. Moody (miner), Hegg (blacksmith), A. Ericson (cook) and A. Hobbs (ass. cook) (Hoel 1966:461). The task was to improve the infrastructure at the site, i.e. to raise a large storehouse, sort and store provisions and equipment, arrange a temporary quay, etc. Mining was not on the agenda. A radio station was established, but it is not known how much it actually

The buildings by Calypsostranda in 1998. Photo: Susan Barr

functioned. In Spitsbergen Radio's logbook for 1911–26, it is noted that the NEC had contact with Spitsbergen Radio on 16 and 18 August 1919, as well as 11 August 1920, the latter signed Bevan, and on 11 August 1920 a telegram signed Luch (Could it be Lash?) was received via a station NEX, which most likely was a call signal for Northern Exploration. This was perhaps the Calypso radio, although another source (P. K. Reymert in *Svalbardposten*, Christmas 2011) gives the Calypso call signal as Cb and a NEC radio at Blomstrandhalvøya as NOX.

Another attempt to sail to Davis Harbour at the end of June was successful, and two men with provisions for six months were left there. In the course of that summer, NEC transported a number of separate groups to various sites on the west and southeast coasts of Spitsbergen to carry out investigations of the various mineral occurrences on which NEC based their claims. Little actual mining work happened however (Hoel:457–464).

At Marble Island the wintering team had sent encouraging reports back to England, and a new team of 14 marble workers was sent from Aberdeen in June 1919. Herbert W. Leech, who led the group, was not particularly satisfied with the winter work. The point of entrance for the workings was not chosen well and preparations for the summer work were useless. In addition, he received the message that the machinery which was due to be shipped up for that summer could not be sent after all. Leech switched plans to one of the Lovénøyane (Lovén Islands – called Breccia Island by the NEC) in Kongsfjorden, where it was possible to dynamite blocks of marble out of the frozen rock face. At the same time other promising marble occurrences in the surrounding area were examined and found to be viable. However, with bad weather and other difficult circumstances, only eight blocks of marble weighing a total of six tons were taken out from Lovénøyane for sale. Even these did not make it to the international markets since the ship which was to carry the blocks to Norway was declared to be in too bad a state to make it across the difficult stretch of sea, and both ship and marble were left for the winter in Kongsfjorden. Before leaving Leech had, however, made sure that the site at Marble Island was prepared for effective working the next year, when he expected 1000 tons of marble could be taken out. A small caretaker force was left there to winter (Hoel:464–65).

The Treaty of Spitsbergen 1920

The NEC had plans for an extensive geological survey along the Bellsund coast and south to the glacier Torellbreen in summer 1920. As they wished to start early in the summer, they had a number of huts built in the area already in summer 1919. Experienced trapper Petter Trondsen, together with Ole Sivertsen, Alf Jakobsen and Ole Bangsund, took on the building job and then used the cabins for winter trapping 1919–20.

Dr Salter had noted in his diary 25 March 1918: "Clinched a deal with the Northern Exploration Co, receiving £1000 for 2100 shares. I think I am wise." (equiv. to £200 000 today). However on 6 February, 1920 he was not so sure: "Spitzbergen property seems to have done well after all. I am the poor chap left out in the cold!"

The geological report for the summer, which covered the various NEC claims regarding zinc, gold, silver, copper, coal and marble along the west coast of Svalbard from Kongsfjorden to Bellsund, and also at Davis Harbour on the southeast coast, concluded with the opinion that only the marble and the Davis Harbour coal could be economically worthwhile. Here, Hoel added the comment in his account that the ice and harbour conditions there made it an unlikely spot for mining (Hoel:467).

On 8 September 1918 Norwegian Prime Minister Christian Michelsen transferred his two cabins dating from 1901 and rights in the area to Alf Frantzen, who was a shareholder in the NEC. One cabin was on the west side of Recherchefjorden and the other below Kolfjellet (Coal Mountain) on the north side of Van Mijenfjorden. At various dates in 1919 Frantzen had other claimed areas transferred or sold to him, including an area around Davis Harbour, an area in southeast Svalbard from Kvalvågen to Edgeøya, an area around Magdalenefjorden, an

The NEC occupations which were registered with the British Foreign Office before 9 February 1920 (Hoel 1966:468). It can be seen that these covered a considerable proportion of the accessible areas of Svalbard!

Okkupasjoner på Spitsbergen, anmeldt til Det britiske utenriksdepartement av The Northern Exploration Co. Ltd. før 9. februar 1920.

area around Trygghamna and Daudmannsøyra and an area around Colesbukta. All these were transferred from Norwegian citizens and notices given to their own Foreign Office – the normal procedure regarding occupations during the *Terra Nullius* period in Svalbard. On 20 October 1920, Frantzen transferred all these rights and occupations he had gained in Svalbard to the NEC: "all my rights of any kind in or on the Spitsbergen Archipelago including Edge Island acquired either in my name or in any other name, together with all buildings tools, machinery, equipment and stores of whatsoever kind" (quoted in Hoel:468).

On 9 February 1920 the Treaty of Spitsbergen was signed in Paris, giving Norway sovereignty over the archipelago. According to the Treaty, companies had to give notice of their claims to the Svalbard Commissioner three months after the Treaty came into force. NEC now had claims covering 10–15 000 km². Hoel notes, however, that there were opposing claims to much of this, including 16 Norwegian claims (Hoel:468). At the Ordinary General Meeting of the NEC in April 1920 *The Times* reported that the news of Norway gaining the sovereignty over Svalbard had come as "a tremendous shock" to the Company directors. They had, however, received a letter from the Foreign Office dated 4 October 1919, which stated that the Company's assets in Svalbard would be undisturbed. It was at this meeting that Ernest Mansfield spoke at some length about his exploring and adventures and about the large mineral wealth and immense possibilities that lay there before them and was described as "the original explorer" (*The Times* 7 April 1920).

For the next five years after the signing of the Treaty, the myriad of, often overlapping, claims to land areas were sorted out and it was no longer possible to claim new areas in the traditional way. Birger Jacobsen switched his attentions to Jan Mayen, an Arctic island further west that did not become a national territory (also that Norwegian) until 1930 (Barr 2003).

The Treaty was finally ratified in 1925, following which a Svalbard Commissioner, Danish legal professor dr. jur. Kristian Sindballe from the University of Copenhagen, was appointed to decide the legitimacy of the various land-ownership claims. Adolf Hoel was appointed by the Norwegian government to assist with this work. Attempts were made to resolve some of the claim conflicts before the deadline for registering them with the Commissioner expired on 14 November 1925. In particular, this applied to the NEC, with their huge claim areas. Hoel's main objection was not, however, the number of km² but the fact that the NEC claimed areas in conflict with the Norwegian coal companies Kings Bay Kul Comp. a/s and a/s Kulspids, and in addition that the NEC claims covered important parts of the coastline in the large fjords on the west of Spitsbergen, where the shipping season was longest and best.

NEC board member Alf Frantzen in Norway had tried in 1922 to obtain recompense for NEC's claim to disputed areas around Kongsfjorden from Kings Bay Kul Comp. a/s across the fjord from Blomstrandhalvøya at Ny-Ålesund. Amongst other actions, British diplomacy had been brought into play and the British minister to Norway Mansfeldt de Cordonnell Findlay had delivered a note on 19 April 1919 to the Norwegian government on behalf of the British government protesting against the fact that Kings Bay Kul Comp. a/s (KBKC) had violated NEC's rights on Spitsbergen. The note stated clearly that the British government would defend the English company's lawfully gained rights in Spitsbergen and support its rejection of claims "founded on acts committed in contravention of the rights of British subjects in these regions" (quoted in Hoel:1145). The British minister suggested that both companies cease working their respective claims until the matter was settled, but this was rejected by KBKC. Talks continued until Frantzen on 14 December 1922 offered to sell all of NEC's claims to the Norwegian government. The reply was that the Norwegian government first had to receive confirmation of the fact that the NEC actually had the most-legitimate claims to their areas above any other claimants.

The discussions dragged slowly on until spring 1925 when Frantzen and the British consul in Norway, Carl S. Sæther, instead addressed their offer to the Norwegian lawyer Carl Lundh, who was chairman of the "process preparation committee" which was preparing the process for all the Norwegian claiming companies. Since they had achieved success in buying out the German interests in Bjørnøya, the Norwegian Foreign Ministry and the process committee agreed to buy out the NEC as well. Hoel states (p.471, 1146) that the NEC wanted to get the business settled before 14 November 1925 in order to avoid the considerable expense from fees involved in notifying such large claims as well as the legal fees that might become necessary to prepare their own process for the Svalbard Commissioner. The Norwegian authorities did not want a legal process over NEC's claims either for many reasons (see Hoel:1147–48) and therefore finally agreed to buy out all the NEC's claims.

All the energy and resources Hoel used to obtain witnesses, as mentioned above, were only partially successful in reducing the NEC's claims, but the NEC – probably pragmatically – agreed in the end that the final claim would be for only 686 km², spread over the whole area, and which were not in conflict with other companies' claims (Barr 2003:129). On 29 October 1925 the Norwegian government agreed to pay £40 000 for all the NEC's claims in Svalbard. Owing to a Russian claim on a piece of the area claimed by KBKC, the NEC agreed to reduce the sum to £37 000 (Hoel:1151).

The NEC is dissolved

NEC's Norwegian agent, Carl S. Sæther, arranged the production of c. 20 tons of zinc ore of excellent quality from Sinkholmen, Bellsund in 1924 – despite the earlier report of this occurrence not being economically viable, and 240 tons in 1925. Other areas were also examined again this summer, with the possibility of both zinc and lead being considered viable. Marble Island, however, does not seem to have been visited again. The expedition in 1926 was the last in NEC's history, when again the zinc occurrence was considered promising, but various problems there and in all the proposed NEC mining sites prevented visible returns for all the efforts involved.

At the AGM in 1925 it was decided that the capital would be reduced by c. £430 000 through a devaluation of the share value from £1 to 2 sh 6 d. At the end of the financial year on 31 October 1926, all the company assets were valued at £163 585 14sh 11d. A balance sheet for NEC dated 31 October 1926 is lodged in Companies House, London (NEC 1926), and was probably the last one produced. It lists assets of just £163 000 despite having previously raised the share capital to £1M. The tangible assets listed at that time include:

Houses and fittings
Plant and machinery
Wireless stations and outfit
Motor boats and lighters
Hospital equipment, scientific instruments, prospectors outfits
Office furniture at London, Tromso and Spitzbergen
Development equipment on Zinc Island and Duck Island

By 1929 the NEC did not have the capital to pay off creditors, so board member Gerald Dudley Smith was appointed trustee to realise the remaining assets (Hoel 474–482).

On 17 January 1928 the NEC's Norwegian agent Carl S. Sæther offered all the company's assets: buildings, machinery and equipment to the Norwegian state for £30 000. Adolf Hoel advised the government that the 58 buildings, which were spread over much of Spitsbergen, would be useful in the future for hunters and trappers to use. The Commissioner of Mines, H. Merckoll advised that there was little chance of developing economically viable mines on the properties. However, this was an important opportunity for more of Svalbard to be owned by the state, rather than to be in foreign hands. After many consultations with various experts, the Norwegian Ministry of Trade offered £5000 for the NEC assets, and refused a new counter-offer of £15 000. After the NEC started to negotiate with a foreign consortium in 1931, the Norwegian government offered NOK 100 000, which was accepted by the NEC in June 1932. Hoel mentions in his account that the government was probably affected by that fact that they had declined in 1931 to buy the Dutch property around Barentsburg, which the Russian government had promptly bought for NOK 3.5 million and still owns today. The monetary figures here are from Hoel 1966:482–487, and unfortunately he switches between £ and kroner. It has been difficult to calculate the exchange rate at that time in order to compare the figures.

The public was informed that the NEC had been dissolved by an announcement in the *London Gazette* on 26 June 1934.

Geologist Adolf Hoel's judgement on the marble occurrences at Blomstrandhalvøya was that although the activities there were the largest in Svalbard after coal mining, there was never an economically positive prospect due to the area's tectonics and the effect of frost on the marble, i.e. that the recurring freeze/thaw cycle cracked the blocks, which then shattered when they were extracted and thawed.

Dr Salter recalled his opinion that at one time the NEC owned most of Spitsbergen, which was recorded in his *Reminiscences* thus:

Now how were we to tackle this business? To take possession of a huge continent almost, like Spitzbergen, was a mighty big order. Did no country own it? Where was Norway? Where was Sweden? Where was Denmark? Where was England? Where was anybody? Surely we could get somebody on proof of value to say, "Yes, we'll own it."

We went to the Ambassadors of the different countries and of Russia – of every country we could – asking them if they would own this. They all said No, it was of no use to them at all. We even went to America, but no country would have anything to do with it.

Now, as I understood it, upon proof that we were the first discoverers we could not be deprived of our primary right to proprietorship, however much we might be in default through not being able to put our hands into our pockets and seek and work the mines. A first discoverer can claim all the while he sits on it. If he relinquishes it his next-door neighbour can go and sit on it, and it may become his. So I had the plans made compact, and I drew up on parchment a statement of what we had done – that we had discovered certain minerals for the first time in the Island of Spitzbergen which were ours by the right of first discovery, the strongest right you can have in regard to any mineral find. I had cases made containing the documents and maps, and got Barclays Bank to put them away in their iron safe in Lombard Street. I then went along to the Foreign Office and saw Lord Percy, Private Secretary to Lord Lansdowne, the Foreign Minister. He and I talked for two or three hours. I said I wanted England to take possession of the whole of Spitzbergen. He said, "I'll do anything I can for you so long as it is not a casus belli – we can't fight about it."

"I understand," I said, "but I want you to take facsimiles of my documents in Barclays and pigeonhole them in the Foreign Office." He said they would do that with pleasure, and it was done.

Some years after we formed a private company to which we subscribed about £9,000, but of course that was not nearly enough. We could put up no more, and so far as we were concerned the whole thing then fizzled out. It then got into the hands of Londoners, Germans, and Americans, who are still making a lot of money out of our original discoveries.

Mansfield, however, did not lose his hold. He made what market he could.

I being the only person who could substantiate the find, on a certain day I went to the cellars of Barclays in London, where the safe was opened by the Sheriff and others wearing chains of office, and we had quite a ceremony of my breaking the seal, showing the contents to those who were going to be interested, and doing it all up again.

Mansfield died in 1923 or 1924, and who is now doing all the big business I know not. All I know now is that if we did anything it would be a casus belli for us! Your name on maps certainly strikes you as being very important at the time, but it does not constitute possession, and you may depend upon it that if we had had any real colours to fly we would have flown them.

The Kongsfjorden area and Ny-Ålesund. Blomstrandhalvøya is off the photo to the left. Photo: Susan Barr

NEC cabins in Svalbard

FOR 25 YEARS FROM 1904, Ernest Mansfield and the Northern Exploration Company made many claims, established various sites and erected numerous huts in pursuit of their mining interests in Spitsbergen. Immediately after the break caused by World War I the NEC stated that their 1918 expedition had provided huts and equipment for 100 men, while the 1919 expedition would "provide accommodation and equipment for at least 500 men" (*The Times* 27.6.1919). The company also claimed to own 5000 square miles of territory by the end of that year (*The Times* 17.12.1919). When the company was sold to the Norwegian government in 1932, NEC was the biggest property owner on Svalbard, owning around 58 buildings, with 34 mining claims spread over at least 16 sites (Hoel 1966:481–83). The maps and summary below give an indication of the scale of the company's operations at their peak.

Many of the cabins were given special "Camp" names by Ernest Mansfield and the Company. However, over time some had more than one name, and the same names were used at more than one location. In addition, over the years some of the sites at remote locations were identified by more than one place name and some huts were even moved to different locations. This results in some uncertainty as to exactly how many NEC huts originally existed and how many of those still exist today.

As well as building their own camps and claim huts, the company also purchased huts from experienced and well-known Svalbard hunters and trappers and employed these men to overwinter at the properties and protect their claims. These included Henry Rudi, Arthur and Johan Oxaas, Gustav Lindquist, Johan Hagerup and August Olafson.

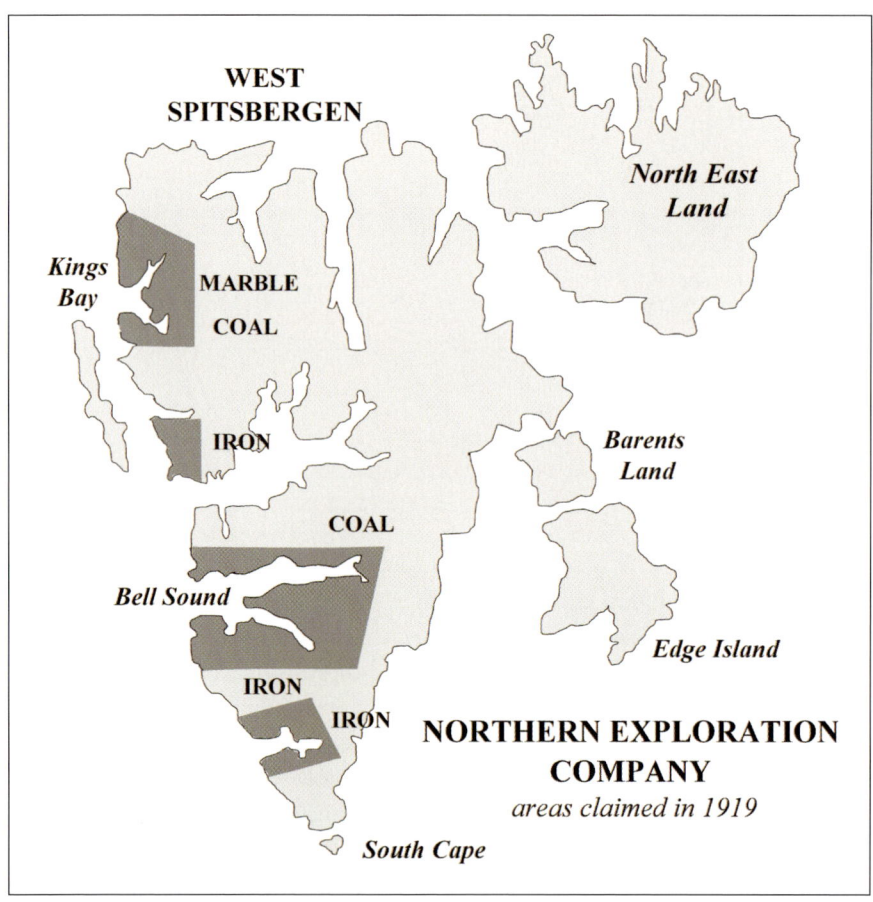

NEC's 1919 claims map, based on a map in The Times of London. Produced by David Newman

All huts and cabins (and all other fixed and movable cultural heritage) in Svalbard that predate 1946 are automatically protected according to the Svalbard Environmental Protection Act (2001). According to this Act "No person may damage, dig up, move, remove, alter, cover up, conceal or disfigure protected structures and sites or movable historical objects, including any security zone, or initiate measures that may entail a risk of this happening" (§42). A total of 162 huts and cabins in various stages from well maintained to ruins are spread around the archipelago, and it is clear that not all of these can be kept in good condition even though the Governor of Svalbard uses considerable resources annually to protect and maintain historical locations. The heritage authorities have therefore roughly divided the huts and cabins into three categories: high-priority objects that shall be maintained at a tradi-

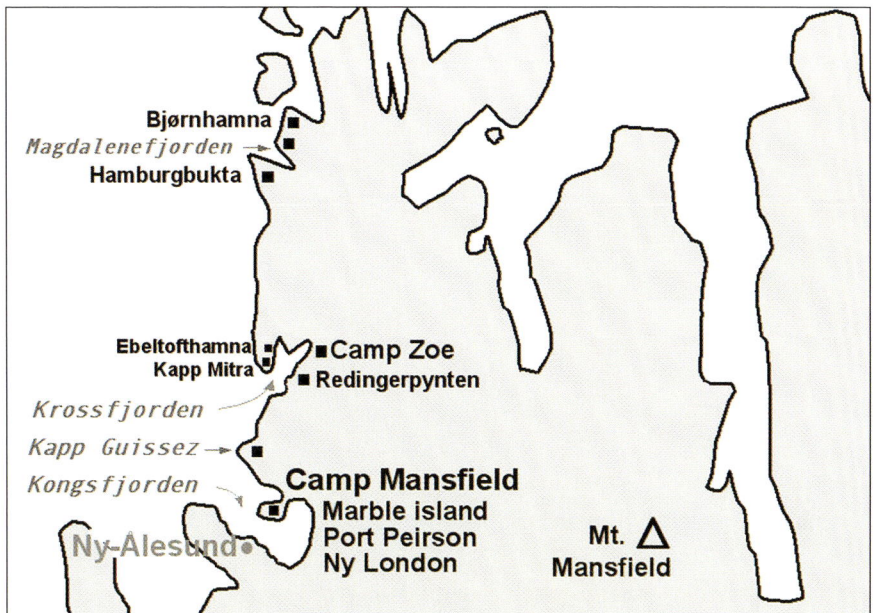

Map of the northwest corner of Spitsbergen. Produced by David Newman

tional standard; deserving objects where the natural degradation is slowed by simple measures such as blocking broken windows or shoring up a leaning wall; objects that will be allowed to decay naturally, but where documentation is secured as early as possible. Many of the NEC huts and cabins are in reasonable condition because they were built in places that have naturally been visited and used for accommodation in the years since. Those in more remote sites are more likely to have fallen into ruin.

The traditional huts and cabins in Svalbard, including NEC ones, were often very simply built: one-layer plank roof and plank walls that were placed directly on the ground and the structure was then covered with asphalt roofing paper. Stones and earth piled around the base of the walls helped to insulate and to anchor the structure to the ground. A door and a window or two, a small oven and bunk bed were basics, and the interior of maybe 6 x 4 m was divided into a small entrance area and a larger living room. The simplicity of the design made the huts easy to erect and to maintain, but they are very susceptible to damage by curious polar bears, which often simply push their way straight through the thin walls, leaving the hut open to more damage from the elements.

The known Camp names, the origin of the names, the place names and locations are presented here in an approximate north-south order. Also included are the company-built and acquired claims-huts that were not given special names.

Bjørnhamna. Photo: Susan Barr 2007

Bjørnhamna – in Northwest Svalbard National Park.
A claim hut was built here in 1912 by August Olafson for the NEC. In 1925 trapper Georg Bjørnnes built a new and better cabin on the site. This cabin has been restored several times and is now kept locked for use by the Governor of Svalbard. It cannot, however, be said to be the NEC hut. Bjørnnes rowed as far as to Virgohamna to fetch materials for his cabin, so he would undoubtedly have used what he could from an earlier hut in Bjørnhamna (Lønø 2002:87).

Magdalenefjord – on the north side of the fjord, in Northwest Svalbard National Park. The hut was built in autumn 1912 by August Olafson for the NEC (Rossnes 1993:120). It still exists but is only just more than a ruin.

The remains of the cabin in Hamburgbukta. Photo: Susan Barr 2007

Hamburgbukta – south of Magdalenefjorden, in Northwest Svalbard National Park.
Built for the NEC by August Olafson in 1912 (Rossnes 1993:120). The hut was photographed standing in 1979 but has now fallen down.

Camp Zoe. Photo: Susan Barr 1985

Camp Zoe at Tinayrebukta, – by Kapp Mitra, Krossfjorden.
Camp Zoe was built in 1911 by Henry Rudi for the NEC and was named after Ernest Mansfield's daughter Zoe Bernice Mansfield who was born in Goldhanger, England, in 1904. It is still kept in good order and is occasionally used by staff from Kings Bay/Ny-Ålesund. There is a

good stream at the site for fresh water. Camp Zoe has been much photographed over the years including a picture in the *Vanity Fair* magazine showing Mansfield standing outside the hut (Vanity Fair 20.8.1913). A short distance away from Camp Zoe are the remains of another possible NEC hut near some signs of marble extractions. There is also a grave here (Avango 2008).

The hut on Redingerpynten. Photo: Elin Lien/ Governor of Svalbard 2010

Redingerpynten – on the east side of Krossfjorden.
A very small claim hut linked with Birger Jacobsen's prospecting activities for the NEC in 1913. It would only be suitable for a single occupant, it is still standing but no longer fit to use. A sign above the door has the name "Nymuen". Quartzite rock can be found in the vicinity.

Kapp Guissez – between Krossfjorden and Kongsfjorden.
Built as a trapping station by Samson Fylkesnes, Karl Finspong and Olaf Rudi (Henry's brother) for their wintering 1910–11. Rudi and Fylkesnes drowned while hunting walrus. Used as a base by August Stenersen and Johan Oxaas in 1911–12 who were engaged by Mansfield as trappers and watchmen for the marble quarry at Blomstrandhalvøya (Lønø 1998:89,100). The hut is still standing and has a painted inscription inside: "Kjøpt av Christoffer Eriksen Eies av Carl S. Sæther for Northern Exploration Co." [Bought by Christoffer Eriksen Owned by Carl S. Sæther for Northern Exploration Co.].

Photograph of Port Peirson from the NEC prospectus. Courtesy of Richard Gardner

Ny-London (New London. The official name is now London, but the name Ny-London sticks) – at Blomstrandhalvøya in Kongsfjorden.
Also known as Marble Island, Port Peirson and Peirson Harbour (Peirsonhamna).
There were at least 10 buildings at this location accommodating 70 men, built for surveying and extracting the marble deposits. NEC's 1913 Marble Island prospectus (NEC 1913) contains a detailed map of the site and many photos. During World War I the site was ransacked and partly destroyed. A large corrugated iron

shed was taken down in 1935 and moved to Ny-Ålesund, where it was used for salt storage and bait herring stocks. Now just two accommodation huts remain, one still clearly exhibiting the "Camp Mansfield" sign, which are maintained and used by the Ny-Ålesund community. The rail route is still visible as are the remains of the workshop and machine shop, a crane at the quay, the remains of channelers, steam boilers and a steam engine. Many of the accommodation huts at the site were given individual camp names by Mansfield (NEC 1913, Booth 1913). It could be that some cabin names were changed according to which NEC supporters were visiting the site.

Camp Mansfield – named after himself (there was also a Camp Mansfield in British Columbia).

Camp Mansfield. Photo: Susan Barr 2008

Camp Peirson, Port Peirson & Peirson Harbour – named after NEC director Sydney Peirson.

The crew with musician in front of Camp Peirson, Ernest in the light suit. Source: The Marble Island Prospectus, courtesy of Richard Gardner

The crew with cook in front of Camp Peirson. Source: The Marble Island Prospectus, courtesy of Richard Gardner

Camp Maples – named after the NEC Secretary J. R. Maples.

Camp Williamson. Source: The Marble Island Prospectus, courtesy of Richard Gardner

Camp Williamson – named after the NEC company chairman Henry Williamson.

Camp Lagercrantz in 2011. Photo: Chris Wainwright

Camp Lagercrantz – after a major shareholder who was also a Swedish government minister.

David Booth in his cabin at Marble Island. Source: David Booth, courtesy of Ross McNeill

Camp Warburg – named after an NEC company director Oscar Warburg. David Booth, Mansfield's brother-in-law and a drilling expert, wrote in his journal during the winter of 1912–13 that his "house" at Camp Warburg was moved several times to be close to the coring machine (Booth 1913).

The four London houses in the foreground at Ny-Ålesund. Photo: Susan Barr

The London houses (London-husene) – Four of the small cabins from Ny-London were moved over the fjord to Ny-Ålesund c. 1950 to serve as family homes for employees of the coal mining company Kings Bay. These four "London houses" have been kept in relatively good condition, and three of them are currently being upgraded for use by visiting researchers. The fourth can also be used but will be kept in the style from the 1950s modifications.

Breccia Island / Maples Island / Juttaholmen – one of the Lovén Islands (Lovénøyane). Named both after the type of rock found there and NEC secretary J. R. Maples. No NEC hut remains on the island, but the foundations of a hut can be seen.

Davis Island – now Storholmen, in Kongsfjorden.
See photograph in the *Ernest in Svalbard* chapter. Built for Ernest Mansfield and named after NEC company director Frederick Davis. Mansfield later referred to the "well built house erected on Davis Island" (Vanity Fair 20.8.1913). The cabin still exists and is situated in an imposing position overlooking Kongsfjorden. Major restoration work was undertaken by the cultural heritage authorities in 2010 and 2011.

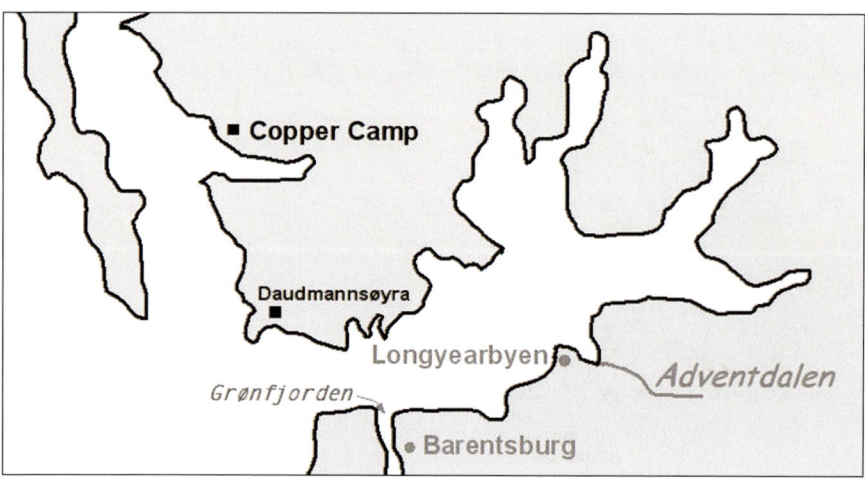

The Isfjorden region. Map produced by David Newman

Camp Copper 2011. Photo: Hilde Tokle Yri/ Governor of Svalbard

The NEC sign on Copper Camp. Photo: Hilde Tokle Yri/Governor of Svalbard

Camp Copper or Copper Camp – St Jonsfjorden, Forlandsundet.

Built by Birger Jacobsen in 1918 for the NEC and named after the mineral found there. The cabin was apparently moved from Camp Smith in Recherchefjorden. Rossnes (1993:98–99) states that the hut was moved again from the south to the north side of St. Jonsfjorden and burned down in 1936. This latter must have been a secondary station as there is still a hut known as Copper Camp standing, although without door and windows. Above the door is the inscription "The southside of St Johns Bay from Cape Müller to Osborne Glacier claimed by NEC Co. London".

Daudmannsøyra – in North Isfjorden National Park.

This site is listed amongst NEC "claims in Svalbard 1927" (Hoel 1966:481). The map published in *The Times* in June 1919 showed a large area of land claimed by NEC in this region (see also Hoel 1966:483). There still is a small claim hut in the area, but it is not known if this was owned by the NEC.

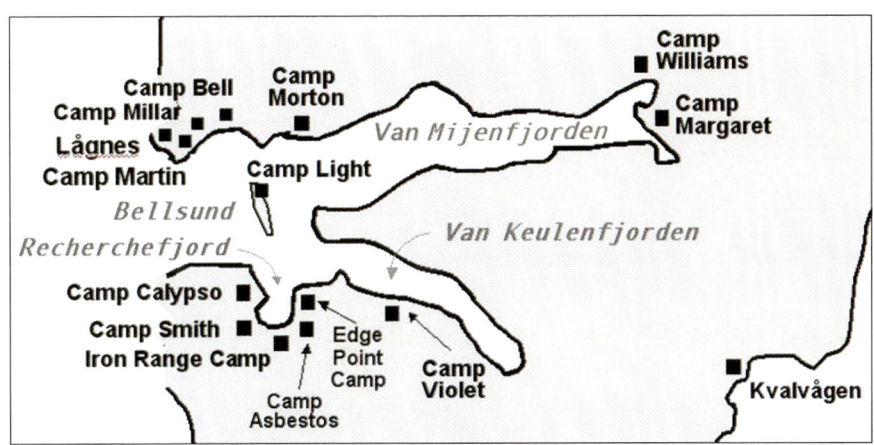

Cabins in the Bellsund region. Map produced by David Newman

Camp Millar 1 and Camp Millar 2. Photos: Governor of Svalbard 2005

Camp Millar – at Ingeborgfjellet in Bellsund.
Originally built in 1908 as a gold prospecting camp, named by Mansfield after NEC major shareholder and mining engineer H. J. Millar. In Adolf Hoel's Svalbards historie 1596–1965 (Hoel 1966:450) there is a photograph of the interior of the "manager house" at Camp Millar where the caption reads "Note the organ on the right". The remains of mining wagons are still nearby and a mine entrance is still visible. The two buildings have recently been restored by the cultural heritage authorities. They are currently borrowed by the Longyearbyen hunting and fishing society (Longyearbyen jeger- og fiskerforening LJFF) with permission from the Governor of Svalbard.

Camp Bell with miners sitting on whale vertebra. Source: Charles Mann collection, courtesy of Rosemary Mann

Camp Bell – at Vårsolbukta in Bellsund, west of the Ingeborg mountains.
Built in 1908 by Charles Mann and George Alexander of Goldhanger (Mann 1908) for the exploration of gold and coal and named after NEC major shareholder Adam Bell. Ernest Mansfield spent the winter of 1908–9 here. Camp Bell has recently been restored by the cultural heritage authorities and has been pulled further in from the shoreline, where it was in danger of being eroded into the sea.

Camp Morton – at Kolfjellet on the north side of the mouth of Van Mijenfjorden.
Originally Christian Michelsen's House, built in 1901. Mansfield and Charles Mann renovated the house in 1908 (Mann 1908) and initially called it Camp Mansfield and later renamed it Camp Morton after the Earl of Morton (1844–1935), an early major shareholder in the NEC and friend of the Rev. Gardner (Douglas 2011). They visited the site in 1906. Michelsen's House was sold to the NEC in 1918, and two prefabricated barracks were built for investigating coal deposits in Kolfjellet, so there

Camp Morton and Clara Ville. Photo: Siri Hoem/Governor of Svalbard 2008

were originally four buildings at this site, including Clara Ville below. Camp Morton is still standing and has been restored by the cultural heritage authorities.
]

Clara Ville – at Camp Morton, Kolfjellet by Van Mijenfjorden.
One of the cabins at Camp Morton, built by Charles Mann and George Alexander of Goldhanger in 1908, and probably named after Charles Mann's relative, his Aunt Clara, with whom he lived in Goldhanger. It is now lent out by the Governor of Svalbard to the Longyearbyen snowmobile club called "To-takteren".

Camp Margaret – south side of Braganzavågen, Sveagruva.
Named after Mansfield's wife, nee Margaret Booth. It is not known how near this camp was to where the Swedish Sveagruva mine was located (Orwin 1939), and there is no evidence that it still stands.

Camp Williams – north side of Braganzavågen, Sveagruva.
Probably named after Vernon Williams, a founder of the NEC in London. Again, it is not known how near this camp was to where the Swedish Sveagruva mine was located (Orwin 1939), and there is no evidence that it still stands.

Camp Williamson – at the mouth of Braganzavågen. Named after Henry Williamson, a director of the NEC. Had disappeared by the end of the 1930s (Place Names). Most probably the same cabin as Camp Williams.

Camp Violet – (Bamsebu) at Ingebrigtsenbukta, east of the Kvitfiskstranda, south side of Van Keulenfjorden.
Named after Mansfield's lover – "a telephonist in an English hotel". It seems that the building was originally an NEC cabin on the east side of Recherchefjorden which was moved to Ingebrigtsenbukta in 1930 for beluga (kvitfisk) hunting. It was then probably much improved. Researchers from the Norwegian Polar Institute named it Bamsebu (Bear hut) many years ago. It is now privately owned.

Camp Gilson – north side of Van Keulenfjorden.
Built by NEC as a simple claim-house and named after NEC mining engineer Kenneth Gilson. Probably disappeared many years ago.

Calypsobyen – Camp Calypso / Calypso City – at Calypsostranda in Recherchefjorden.

See photograph in the *Ernest in Svalbard* chapter. Together with the Marble Island/Ny-London site, this is the largest of the NEC camps in Svalbard, mostly established in 1918–19. The coal occurrence was small and the mining activity at the site was mainly over in 1920. There were originally eight buildings at the site, the oldest from the Christian Michelsen expedition in 1901. A long living barrack and a mess barrack were erected for the NEC in 1918–19, as well as a radio station from 1919, which was probably very little in use. There are six remaining buildings, one of which – originally the mess barrack – has been used regularly by Polish scientists since 1986. Blomlihytta, built for winter trapping by Birger Jacobsen in 1919, was restored by the cultural heritage authorities in 2010 and other restoration work was done at the locality in the 1990s. Remains from the mining period lie around the area and include a rail track with wagons and several boats (Governor of Svalbard reports).

Camp Jacobsen / Strandhuset / Michelsenhuset – one of the 8 buildings at Calypsobyen in Recherchefjorden and the oldest. It was built in 1901 for Christian Michelsen (later to be Norwegian prime minister) and sold to the NEC in 1918. Named by the NEC after Birger Jacobsen, the NEC site manager who sold all his claims and huts to the Company.

Iron Mountain – Camp Jernfjellet. Photo: Hilde Tokle Yri/Governor of Svalbard 2010

Iron Mountain Camp / Iron Range Camp / Jarnfjellet – east side of Recherchefjorden.

Named by the NEC after the ore found in the nearby Martin Mountains. There were originally four buildings. In 1936 there were only three and by 1984 only one skewed building remained (Rossnes 1993:52), which is now in a precarious state.

Camp Smith / Giæverhuset – west side of Recherchefjorden.

See photo in *Ernest in Svalbard* chapter The house was built in 1904 by John Giæver of Tromsø as a summer house. It was sold to the NEC in 1911 and named after Gerard Dudley Smith, a director of the NEC and of the Union of London and Smiths Bank. The building still stands, albeit leaning heavily.

Camp Asbestos – Asbestosodden, west side of Recherchefjorden.
There were claim huts from both the NEC and a/s Kulspids here, the latter from 1918 and the former built by Carl Sæther, NEC's Norwegian agent, and Birger Jacobsen in 1921. Asbestos occurrences can be seen on the surface. A second NEC hut was apparently erected a little further west. The huts are now in ruins.

Camp Volage – east side of Recherchefjorden near Reinodden.
The site of both English seventeenth century whaling and later Russian winter hunting. The NEC apparently erected some buildings here, but none remained in 1936 (Rossnes 1993:52).

Camp Light – on Collinderodden, south side of Van Mijenfjorden.
Johan Hagerup sold this hut to the NEC. It was named after Charles Light, an NEC office manager. By 1936 it had disappeared.

Edge Point Camp – at Lægerneset, in Recherchefjorden.
Named after the English whaler Thomas Edge (1587–1624); therefore, it is unlikely that NEC gave the name. There are traces of NEC activity, but no building remains (Askeladden).

Kvalvågen – on the west coast of Storfjorden, south of Kvalhovden.
According to Carl Sæther this was built as a claim hut for the NEC in 1923. It stood in 1936, but it no longer exists (Governor of Svalbard).

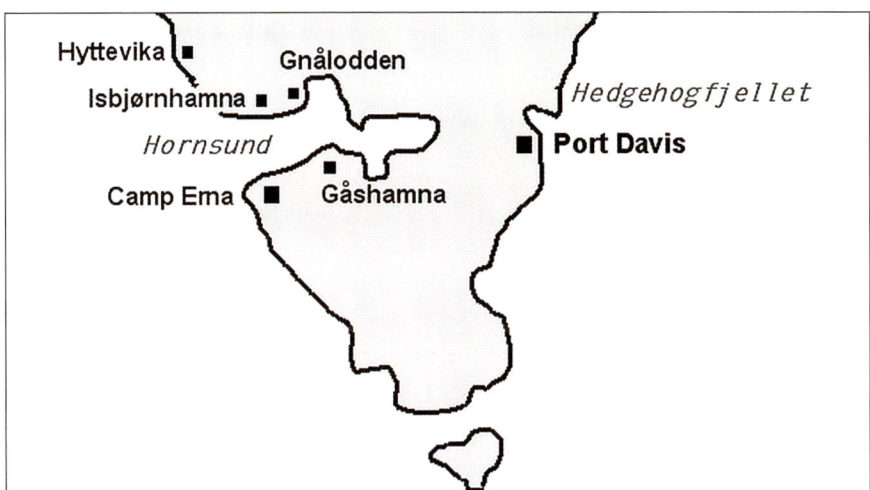

Cabins in southern Spitsbergen. Map produced by David Newman

Isbjørnhamna – near Gnålodden, Nord Hornsund.
Built in 1919 by Anders Kvive Andersen of Tromsø and Ture Lifbom of Stockholm who were wintering trappers for Johan Hagerup, then sold by Hagerup together with ten other huts to the NEC. There was an NEC ownership plate at this site in 1936, and the name Hagerup was written on the door (Rossnes 1993:41). There have been other huts in the area, and it is possibly another one that is now used by scientists from the Hornsund Polish research station.

Camp Erna – Gåshamna, south Hornsund.
Named after Erna Gurine, daughter of Gustav Lindquist, a well-known Svalbard trapper & NEC employee, who built the hut in 1919 for the NEC. It was originally called "Camp Lindquist". It is believed to have disappeared now.

Port Davis / Davis Harbour – Hedgehog, by Storfjorden, on Spitsbergen's east coast.
Named after NEC company director Frederick L. Davis and still known as Davishamna, although there is no harbour there now. NEC employees wintered there as guards for the Company in 1920–21, 1921–22 and 1923–24. There were originally three buildings at the site and the current state is not known, although it can be presumed that there are no standing buildings.

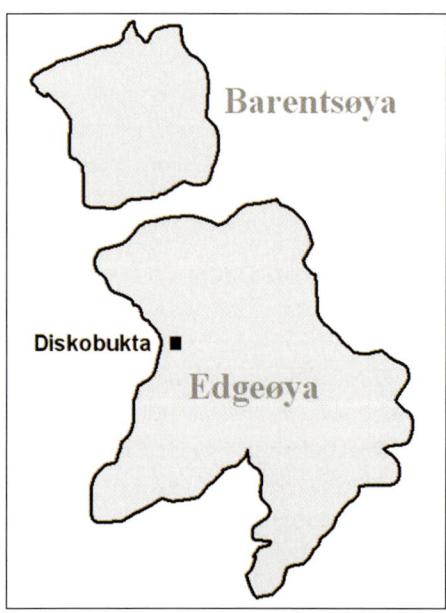

Diskobukta – on the west coast of Edgeøya (Edge Island).
Carl Sæther identified a hut at this location as an NEC claim hut. This has now disappeared (Governor of Svalbard).

Map of Diskobukta, Edgeøya, produced by David Newman

A Summary

Thirty-one Mansfield and NEC named camps spread over 33 sites have been identified and are mainly along the west coast of Spitsbergen. The locations of approximately 57 NEC-owned buildings in total have been identified, and around 27 of these are still standing. In 1928 Adolf Hoel advised the relevant Ministry (Ministry of Trade) that they should include the NEC's 58 buildings in the agreement to buy the NEC's assets in Svalbard. The deal was clinched in 1932 (Hoel 1966:483).

There were, and still are, many other huts in Svalbard not associated with NEC that were built by hunters, trappers, other explorers and mining companies. The cultural heritage authorities' database of historical sites in Svalbard has registered c. 230 huts/cabins defined as "trappers' huts" and remains of such in the archipelago.

Location of NEC cabins in Svalbard. Map produced by David Newman

About 162 of these are huts/cabins in various degrees of still standing, while 71 are visible remains in the ground of what once were huts (Askeladden). The numbers are not exact, since the objects have not always been easy to identify. Using the number of 230 historic huts on the archipelago, it would seem that the NEC was responsible in some way for about a quarter of the recorded historic huts. Many of NEC's huts were also used for winter trapping even while the company was in operation, serving a dual role of being a claim hut, manned over the winter by NEC-employed trappers to protect the claims, while these men supplemented their income and their provisions by hunting and trapping the wildlife.

All of NEC's huts were located close to the coast in fjords and natural harbours giving easy access from the sea. Today a number of these huts can still be used, but the use is controlled by the Governor.

Ernest's Literary
and musical endeavours

Ernest's literary and musical creativity evolved and developed throughout his career and encompassed (in approximate chronological order) short story and poetry writing, music and composing, editing magazines, writing newspaper and promotional articles and finally the authorship of two novels.

Short Stories

Ernest's first known literary endeavours were in 1889 at the age of 27 while living in Wanganui, on the North Island of New Zealand. A literary magazine called *The Family Friend* (10.7.1889) published two of his short stories entitled *Father's Picture* and *Three Christmas Days*. Later in that same year the *Canterbury Star* (19.11.1889) newspaper published anonymously a story entitled *A Terrible Ride*, which a year later was reprinted in a supplement of the *Wanganui Herald* (31.10.1890) under his name. The story is very short and describes an incident and an apparent murder on a stagecoach going to Glasgow.

Articles in the *Wanganui Herald* (1890) over the following year refer to seven of his short stories:

Two Chums, Three Christmas Days, Father's Pictures, A Terrible Ride, A Poor Little Waife, How I Got My Name and *A Real Romance*. The *Wanganui Herald* (15.11.1890) reviewed *Two Chums* thus: "… Mr Ernest Mansfield, whose literary efforts are always perused with interest. The plot of "Two Chums" is of a most romantic and engrossing description. A number of the most interesting scenes being laid in New Zealand, while the characters give excellent scope for some intensely dramatic situations".

In January 1891, *How I Got My Name* was printed in full in the *Wanganui Herald* (17.1.1891) and was sub-titled: *A true and authentic account of a christening party*. The story is based on a journey on the London South Western Express train.

A Real Romance was also printed in June 1891 in the *Wanganui Herald* (16.6.1891). The newspaper introduced the story with "On our fourth page will be found an interesting and locally-written story, entitled, "A Real Romance", by Mr Ernest Mansfield, who will be remembered as the author of "Three Christmas Days" and several other excellent literary efforts". Only the first scene has been found and is about a miserly father and his desire to find a suitably rich husband for his beautiful daughter called Zoe. Ernest named his daughter Zoe in 1904, some 13 years after this story was written.

Poetry

Ernest wrote poetry both at the beginning and end of his career. In August and September 1891 while Ernest was 29 years old and was the Garrison Band secretary, adverts for social events in the Garrison Drill Hall appeared in Wanganui newspapers that included his name and poems which were most probably written by him.

There is a poem on page 151 of Ernest's first novel *Astria – The Ice Maiden* written in 1910 entitled: *O' Brightly Shines Aurora's Beams* and headed:

TO AURORA

O' Brightly shines Aurora's beams
To illumine the skies.
The glory of the glowing gleams
Is the love in her eyes.
The streamlets of delicious light
Are her tokens of grace –
The dazzling wavelets on the night
Are the smiles on her face.
Those silv'y waves of beauteous splendour,
Lighting up the land and sea,
Are messages so true and tender
In sweet *billet doux* to me.
Bewitching queen of night – O, I adore her.
A Goddess of delight – my sweet Aurora !
My beautiful, my sweet Aurora !
Thy caresses flow above
Are softer than sighs,
Thy radiance is the Lamp of Love
Which beams in thine eyes !

Wanganui Chronicle 26 Sep 1891

WANGANUI GARRISON RINK DRILL HALL.

Hats and Caps, and Caps and Hats,
   Methinks I hear you say why that's
A strange thing indeed to give a prize
   Of HALF-A-GUINEA for the best largest size;
And the best of it is, that that is'nt all
   For the same amount you give for the small.
And the very best race of late that we've made is
   The one set apart for the dear little LADIES.
And another great prize, I must tell you all that.
   A GUINEA we give for the most novel Hat.
And a Gentleman's Race is too on the board,
   So sweet pretty GRACE take dear little MAUD;
And all you young Ladies should not fail to go
   With your sweethearts and friends to this our last Show;
When with strict truth it may fairly be reckoned
   The GREATEST of CARNIVALS on OCTOBER the 2ND.

GRAND HAT AND CAP CARNIVAL,
AND
LAST NIGHT OF THE SEASON.

FRIDAY, 2ND OCTOBER.

Half-guinea Prize for best and smallest hat.

Half-guinea Prize for best and largest hat.

And One Guinea for the most Novel Hat.

The public to decide the winner.
Half-mile Handicap Flat Race for Gentleman Skaters.

And a Ladies' Race if a sufficient number of entries are received.

For the Ladies' Race a very handsome present will be given to the winner, and no charge for entries.

Ladies intending to contest in this event please give their names to
   F. J. TASKER,
      Manager.
Or   ERNEST MANSFIELD,
      Hon. Secretary.

One of Ernest's poems. Source: Wanganui Chronicle 26 September 1891

The following poem was written by him and placed on a trapper's grave in Svalbard in 1911 and was later recorded in Birger Jacobsen's obituary of Ernest Mansfield (Oslo Aftenavis 16.1.1925):

> Slowly and sadly we laid him down,
> From the field of his fame fresh and gory;
> We carved not a line – we raised not a stone,
> But left him alone in his glory.

Carl S. Sæther, who had considerable interests in both mining and trapping in Svalbard and from 1920 became the NEC's local representative in Tromsø, wrote an article about Ernest Mansfield that was published in the newspaper *Tromsø* (Sæther 9.4.1934). In this he wrote that he found a poem in Ernest's diary after Ernest's death. It has been translated into Norwegian and now back again into English and has suffered in the process but is well worth including here:

> What do I care if a hurricane can come snorting
> Around my little cabin by the frozen sea
> And if a bitter snowstorm now comes freezingly
> It will only be the desert that winter gave me.
> It is just joke and fun now with wild storms
> And 50 degrees of frost causes no sorrow.
> Today I am so happy and glad for all forms of life:
> For I know that the sun comes home again tomorrow.
> There will be a party if the cloud clears
> So old sun can show itself, the light king,
> To be able to see its happy rays after long winter nights
> To be able to heat the marrow in my frozen body.
> It may be that dark clouds will stop.
> Well, then I take from the imagination a little guarantee,
> And believe that the cloud is a lively blue, with white edges.
> For I know that the Sun comes home again tomorrow.

Music and Composing

Ernest first advertised banjo lessons in the *Wanganui Herald* (9.1890) when he was 28 years old: "Mr. Ernest Mansfield has made arrangements to give instructions to all who wish to learn, the art of tum tumming." He was said to have been "a pupil under Mr. Herbert J. Ellis, the celebrated London player and composer, and consequently his teaching is of a first class order."

In October 1890 the *Wanganui Chronicle* (20.10.1890) reported that Ernest played the banjo at a Garrison Band Concert and sang *The Golden Slipper* with a banjo accompaniment. At the same concert he also supplied the music for the American clog dance. In January 1891 it was reported that he sang *There's Gold in the Mountains* at a concert, and in August 1892 he performed *Waltzing Round the Waterbutt* at a Maxwell Town concert.

In the late 1880s and early 1890s, the name of E. Mansfield appears many times in Wanganui newspapers in association with Garrison Band concerts and activities.

Between August 1892 and January 1893, Ernest extensively advertised a song he had composed entitled: *The Wheel's the Life for Me*. In September 1892 *The Otago Witness* (1.8.1892) wrote: "I have to acknowledge the receipt of a song, "The wheel's the life for me," from Messrs C. Begg and Co. The song is eminently suited to wheelmen, and is likely to become popular as a cycling song. The air is tuneful and of easy compass; and there's a fine swing about the chorus which given with full force from the lungs of a body of cyclists out on tour should rouse the enthusiasm of the most despondent rider. The song is written and composed by Mr Ernest Mansfield, of Wanganui".

The *Wanganui Herald* (5.1.1893) reported: "The *London Referee* of the 17th of October last has a flattering notice of Mr Ernest Mansfield's bicycle song. Seeing that the writer is that sterling dramatist and author, G. R. Sims, the compliment may be said come from one of the best authorities in England. It is also pleasing to learn that the song will be published in England, and we shall not be surprised to see it have a great run. Thus says Mr G. R. Sims: "All the way from Wanganui, New Zealand, Mr Ernest Mansfield sends me a capital new bicycle song and chorus, entitled – The Wheel's The Life for Me – which he has composed. I have tried it on my trike round the inner circle and cordially recommend it. I should strongly advise any bicyclist who may take Wanganui on his way, not to forget to secure a copy". Sadly no copies of the music score have been found in either New Zealand or England.

Nine years later in British Columbia *The Kootenaian* (9.11.1899) reported: "Music in the Air at Camp Mansfield. Ernest Mansfield sends instruments to the camp for the winter".

A reference to Ernest's musical interests can be found in a long article in a 1911 *The New York Times* (13.8.1911) article entitled: "Village Trio Seek Spitsbergen Gold". Here Ernest is described as "a musician". Very similar articles with the same phrase were published around the world.

Finally, his composing talents are demonstrated in his second novel *Ralph Raymond* published in 1913, which has two songs complete with music scores and lyrics on pages 203 and 340.

Magazine articles

Between 1893 and 1895 Ernest embarked on a role of author, editor and publisher of several magazines, with some of the magazines it would seem being produced simultaneously. In 1893 the *Wanganui Chronicle* (7.1.1893) reported on a: "new weekly magazine entitled *The Stage*, which is "to be introduced to readers on this coast under the conductorship of Mr Ernest Mansfield. It will be devoted to "music, sport, and the drama," and its readers are promised notes and paragraphs, as short and crisp as they can be made, on all subjects. The ladies are not to be forgotten, as stories and special notes for themselves are to find a place in its pages. Brevity is to be aimed at in all things, and the first issue will consist of four pages only. We understand the paper is a private venture, and that a cheap pennyworth is the proprietor's main idea". While in the same week the *Wanganui Herald* (9.1.1893) wrote: "Mr Mansfield has written largely for the London papers, has contributed to the leading Australian journals and supplied a great deal of reading matter for the principal papers in both islands of New Zealand. It is easily seen that the conductor is no novice at wielding a pen".

In September of that same year, the *Wanganui Chronicle* (1.9.1893) reported that Mr E. Mansfield's little theatrical publication "The Curtain" contained the cast of characters of "The Broken Idol" and was published as an insert to the newspaper. It contained a competition, and the newspaper later reported that 1500 entries were received. Soon after, in November of 1893, the *Wanganui Herald* (7.11.1893) reported: "We have to acknowledge the receipt of the first number of *Fair Play*, a new weekly illustrated journal published in Wellington... Mr. Ernest Mansfield, formerly of Wanganui, is acting as travelling representative for the new journal, and will no doubt get a large number of subscribers during his peregrinations".

Ernest's name only appears in the magazine to declare that he was their "travelling representative", with no clues as to who edited or wrote any of the articles or how many Ernest might have written. Most articles do not identify the author, and many articles give as "from our correspondent". An article entitled "Music in Wanganui" could well have been written by Ernest, although there is nothing specific to attribute it to him. All 27 issues of Fair Play can be viewed at *www.paperspast.natlib.govt.nz*

In 1894 Ernest was involved in yet another publication entitled "A1". This time he was identified as the publisher of the periodical, which produced 77 editions between 1894 and 1896. The *Wanganui Chronicle* (13.7.1894) referred to it as: "our new A1 supplement" and it appears also to have gone under the heading of: "Gems of thought from noble thinkers". The newspaper quoted: "Wisdom and humour will go hand in hand and the reader, tiring of a scientific treatise or a social sketch, will be able to turn to the light and airy writings of the humorist, or rather humorists, and revel in jokes innumerable. Practical and useful knowledge, natural history, social and general news, fact, fiction, prose and poetry, indeed every scrap of reading matter of real worth and genuine interest that the compilers can lay their hands on will be appropriated and reprinted for the benefit of our subscribers. … Although the initial number will be published without the instalment of a serial story, the management have made arrangements whereby at an early date this omission will be rectified and our readers supplied with a serial from the pen of one of the most eminent writers of the day…". One can only speculate as to who might have written these words.

Dr Hatherley is identified as the editor of *A1* and Ernest was the publisher; the relationship between these two men was to play an important part in Ernest's career over the following three years.

In February 1895 the *Wanganui Chronicle* (28.2.1895) reported on an A1 Company shareholders' meeting and recorded Ernest as a director and the chairman of the A1 Company. Later in that year the same newspaper (25.4.1895) reported on a "Museum Exhibition and Fancy Fair": "No small attention was devoted to the perusal of the first issue of the *Wanganui Museum Gazette*, a neat little journal published by Mr Mansfield for the Committee. The Gazette will be issued every evening" – a very ambitious and perhaps unrealistic undertaking.

Five years later, when Ernest was in British Columbia, a notable seven-page article appeared in *The Canadian Magazine* (11.1901) entitled: "A New Canadian Glacier" and subtitled: "The Story of Mountain Climbing in the Rockies" — despite the fact the glacier was not actually in the Rockies. The article was written by Welford Beaton, who was editor of the *Nelson Daily Miner* between 1901 and 1902, and had earlier been one of Ernest's business partners and his company secretary. As Ernest features so much in the article, one can only assume he had a large part in the authorship and also that Mr Beaton had a very considerable respect for him. Here is part of the introduction:

> One of the first men to cross [the glacier] was Mr. Ernest Mansfield, a mining engineer, who represents English capital that is looking for dividends in the wealth-laden mountains of the Kootenays. He had secured a property late in the season and a sudden and heavy fall of snow had cut off his only known means of access. But he tackled it from another direction, and after very nearly losing his life, a fate which his two companions would have shared, he got safety over the great glacier and returned to his mining' camp long after the other members of the party had quite decided that they were lost. The glowing description which Mr. Mansfield gave of the beauties of the glacier, which he christened the Kitchener, quite determined me that not another season would pass until I had climbed to its very summit.

In 1913 a five-page article appeared in the *Vanity Fair* magazine (20.8.1913) written by "our special correspondent in Tromsø Ernest Mansfield" and was entitled: "A Struggle with Death in the Arctic". It described in some detail how NEC employees based at the marble quarry on Blomstrandhalvøya participated in a rescue mission with an expedition to find the lost German expedition led by Herbert Schröder-Stranz. The episode is described in the *Ernest in Svalbard* chapter, *Winter 1912–13*, but an extract from the article is included here since it demonstrates Ernest's experience and skill as a reporter:

> In summer this is the beauty spot of the world, surrounded by majestic mountains, and mighty glaciers, amidst a grandeur impossible to excel; the deep blue waters of the Bay flowing idly over polished marble shores, human-eyed seals disporting themselves in beautiful ornamental marble basins, alongside lovely moss-lined beaches, and lady eider nesting everywhere. But it is winter. King's Bay is a sheet of ice, and locked in its embrace is Davis Island, covered with a seven-foot mantle of snow. When the men had halted they decided to change one of the sleighs because of its weight, and Dr. Wagener and his companions returned to Cross Bay for a lighter one. This was done, and the real start was made the next morning.
>
> Great difficulty was experienced in scaling the Kings Glacier, but a passage was at last found on the south side. The treacherous ice ridges and deep chasms were safely crossed after a hard day's march; and they arrived at the first crown where a depot was fixed – a distance of twenty miles from Port Peirson. James Booth, who had the lead in Herbert Millar's sleigh, fell through a snow bridge down a crevice in the ice, but fortunately he was roped, and thus escaped death, for at the bottom of the abyss are torrents of water which are continually running, winter and summer, and tearing this gigantic icefield down by the million of tons every day.

The article reported that the author himself was returning the rescued men to Tromsø in the NEC ship *Activ*. This incident and presumably the article resulted in the German authorities of the day naming a mountain on Spitsbergen after Ernest.

Journalism and News Briefings

The Wanganui Electoral Roll in 1893 listed Ernest's occupation as "journalist", but as he turned his attentions towards gold prospecting and the associated stocks and shares trading in the mid-1890s so his literary skills increasingly turned towards publicity and apparently the circulation of newspaper "briefings". Undoubtedly, his earlier experiences as a magazine editor and newspaper reporter had taught him the importance of news briefings within the newspaper industry and business generally, and he had acquired the first-hand experience of how to handle them. A hint of the source of an anonymous briefing is given in this extract from the *Wanganui Chronicle* (8.7.1895): "The following wire was received on Saturday night by a Wanganui gentleman from the broker to the company…". It is partly due to the wealth of newspaper reports and articles that have been identified more than one hundred years later that we can now piece together Ernest's biography.

A series of very optimistic, but unattributed, reports of gold finds at the Lydia mine in New Zealand newspapers involving Dr Hatherley and Ernest were followed by extensive local newspaper advertising of shares in the mine, with Ernest Mansfield identified as the broker for shares in the "No Liability Company". Later a letter to the newspaper questioned the legality of the "No Liability Company".

In total, some thirty newspaper reports, editorials, articles and readers' letters appeared in mid-1895 in relation to the Lydia mine developments.

In mid-1897 Ernest suddenly left New Zealand and while in the UK attended a Diamond Jubilee celebration in London. A long and impressive report "from a London correspondent" was sent to the *Wanganui Chronicle* describing the celebrations and reporting on a speech given by Ernest at a banquet for Australasians. He later established himself in Nelson, British Columbia, from where in 1899 onwards a series of articles describing his activities continued to appear in the Wanganui newspapers. Ernest himself was the most likely sources of these reports. In total over 90 articles referring to his activities in New Zealand and Canada have been identified in New Zealand newspapers.

At the same time that newspaper articles about Ernest's activities in British Columbia were still appearing in the Wanganui newspapers, very similar optimistic reports of his mining activities were beginning to appear in British Columbia newspapers, such as the Nelson *Tribune*, *The Kootenaian* and the New Denver *Ledge*. Perhaps one of the most telling episodes of that period that demonstrates Ernest's character and ability to exploit the newspaper media through briefings was the occasion in October 1900 when he was jailed for a week in Nelson. Over that week there were at least eight long newspaper articles about his plight, all showing signs of his characteristic writing style, and all probably written by him, or at least heavily inspired. Ernest had not been convicted of any criminal activity; rather he was jailed at the apparent instigation of his own Camp Mansfield workers because they had not been paid. The newspaper articles during his incarceration explained that he had made his best endeavours to acquire more funds from their French owners without success. Today it is clear from the articles and his subsequent release that this was in fact a ruse orchestrated by him and that in all probability he was responsible for the extensive press briefings published at the time. He most likely spent most of the time in jail writing the articles.

After his release, press reports of the activities at Camp Mansfield continued as before. Associated with these public relations exercises were some professionally produced photographs of his camp taken by photographers Wadds Bros. of Nelson, which are still in the possession of members of Charles Mann's family in the UK. All the pictures have Ernest himself strategically situated in the centre of the pictures. In total over 110 newspaper and magazine articles referring to Ernest's activities have been identified from his period in British Columbia.

During the European phase of Ernest's career, fewer newspaper and magazine articles seem to have appeared, perhaps due to the remoteness of Svalbard at the time, or perhaps because he began to concentrate on writing his books and the more direct publicity associated with his company, such as the two prospectuses. In 1910 an article appeared in the *London Times* (10.10.1910) under the heading of "An Arctic Adventure" describing an incident where Ernest saved a Norwegian ship and rescued its crew. Two weeks later a similar, but longer, article appeared in the *Wanganui Chronicle* (24.10.1910) under the same heading. One can only assume that Ernest had a large hand in writing these articles and persuading the newspapers to publish them.

In 1911, some thirteen years after Ernest had left New Zealand, a lengthy and impressive article containing the "Gold, or I'm a Dutchman" quote and most probably written by him, appeared in various major newspapers around the world including: the UK *Daily Chronicle*, the *New York Times*, the *Wanganui Chronicle* (NZ), the *Grey River Argus* (NZ), the *Western Mail*, Perth, Western Australia, *Barrier Miner*, New South Wales, the *San Francisco Call* and others. The *Grey River Argus* (Argus 27.9.1911) version is included in the *Ernest in England* chapter. The article appeared in at least seven different newspapers on three continents over the period of a month.

During the European phase of his career, some 30 newspaper articles have been identified in newspapers of the period, mainly in the *London Times* and all referring to NEC company affairs. However, perhaps one of the most notable news articles during the European phase was the *Vanity Fair* article (20.8.1913) described in the section on magazines previously.

In the winter of 1913–14 Ernest revisited Western Australia where he had briefly operated in 1896, and an article in the Kalgoorlie *Western Argus* newspaper (20.1.1914) reported: "…a fine raconteur and tells thrilling stories of his experiences in the ice and snow…".

Ernest's two books

In 1904 Ernest moved to Goldhanger, Essex, UK and there established a friendship with the local rector and doctor. For Ernest an association with these two people must have been extremely fortuitous, as both were wealthy and well travelled. The doctor was also a prolific writer and maintained a detailed diary of all his endeavours throughout his life. We know that Ernest also maintained a diary covering the periods of his lengthy stays on Spitsbergen, but the whereabouts of his original diaries are not known.

Astria – The Ice Maiden
While overwintering alone in Spitsbergen during the winter of 1908–09, Ernest wrote the first of his two books, entitled *Astria –The Ice Maiden* (Mansfield 1911). The book is a semi-autobiographical science fiction based novel and was published as a paperback in London by The Lonsdale Press in 1910. However, the copy in the possession of one of the authors has been especially bound in mock leather to make it into a hardback. The copy was found in 2005 in a salesroom not many miles from Dr Salter's home in Essex together with a copy of *Ralph Raymond*. Both books had been personally annotated, probably by Ernest's daughter Zoe, and include original photographs of Ernest's family affixed to the fly sheets. It is most probable that the books originated from Dr Salter's house.

The book has 157 pages of text with no illustrations. The print is large and well spaced, so this book is short in comparison with Ernest's second novel *Ralph Raymond* and may be typical of a low-cost paperback at the time. It begins with a very generous dedication to Dr Salter, explaining that the author had taken the doctor's advice on the style of the book. Ernest also reveals that the principal character *Daric Clittiem* "somewhat reminds me of a man I always met at your home", so it is clearly based on himself. The dedication neatly complements Dr Salter's praise for Ernest in his diary *Reminiscences*. An excerpt follows and shows Ernest's characteristic flowery style:

Ernest's wife and daughter. Photo in the current authors' copy of "Astria"

Cover of the "Astria" book in the current authors' possession

Dedication

To Dr. J. H. SALTER

[...]

What a number of experiences I have had since [our first meeting]! Some were awful, others terrifying! There was the time when the waves of a maddened sea surrounded me, and threatened to sweep away my house and all my 'belongings—Aye ! and me too! For three long days and nights the seething foam of a furious sea swept around my home continually, surging up the banks of my lonely shack, pouring its waters through the windows, saturating everything! During the storm everything was black—except the sea. That was white! Giant waves were lashed into a milky fury, making a noise like cackling thunder as the crests curled over the waters, and boomed like guns as mountainous volumes struck the shore! It rained, hailed, and snowed the whole of the time. I had to flee from the house again, and again, fearing every moment all would be swept away. In doing so, I had to wade through the flying surf up to my knees. When in the open, the elements were stinging, and I was forced to seek the shelter of my wave-washed house, to escape being frozen to death. Oh! how I longed for some companion! But I was in a lonely world, with everything silent, except the roaring, raging sea!

But winds wilder than the waves were mad. They battered and threatened my staunch little shanty all through the long Arctic night. These were more trying than the troubled waters ! For ten months I was alone ! No one to speak to—not a soul! Ten months! It is a long time to live in solitude, and in that frozen world where I was the only inhabitant, every moment seemed to be accentuated. There were times when I would willingly have given five years of my life, for the pleasure of ten minutes conversation with a Britisher. But such a boon was impossible. I was cut off from the world; with no hope of a word from anyone, or news from anywhere. Absolutely alone, in an uninhabited country, frozen harder than adamant!

[...]

I Remain

My Dear Doctor,

Yours Gratefully and Faithfully,

E.M.

Tolleshunt D'Arcy, Essex.

The first two chapters of the novel are clearly autobiographical and would appear to be a factual recollection of both Ernest's early life and initial experiences on Spitsbergen in the real locations of Advent Dale and Advent Bay (Adventdalen and Adventfjorden). For example: "I went to Spitzbergen with an English clergyman" – a reference to the Rev. Frederick Gardner. "I telegraphed for an Englishman to come at once. He was a good all round man, just the sort to help me lick a camp into shape", and later: "I'll stay if you like Mr. Clittiem," was the last appeal of my man. "I don't like leaving you alone like this!" "I'm all right Charlie," I replied. These two extracts undoubtedly refer to Charles Mann, Goldhanger builder and carpenter, who accompanied Ernest on two trips and was employed to build cabins.

A paragraph in the second chapter must be one of Ernest's most notable pieces of prose:

> Time flies even in Spitzbergen! The 25th day of October arrived, and still no ship had called. It had been a cold day, the country was all white, and there was a hard bone in the ground, which would not get soft again till next June. The sky was free from clouds, in the South there was a kind of twilight. I knew this was the last time Sol would call this year, so I hoisted my Union Jack to greet him. Then, at a few minutes before mid-day, slowly and majestically, the golden orb arose from the sea, hovered for a few minutes above its surface, then slowly sank. When the gleaming edge of the blazing globe sank from view, the waters turned red, the ice lined shore purple, whilst the snow on the mountains assumed every colour of the rainbow. "Good-bye" I cried to the setting sun, when a breeze caught the folds of my flag, and waved it towards the god of life, who had now gone to gladden the homes of Southern latitudes. I hauled my flag down and kissed it. I should not want it this year—that was certain! So there we were; the flag, myself, and the shanty! We shouldn't see the sun again till the 18th of next February, and then only for a few minutes, if the sky should be cloudless!

The "Author notes" at the end of the book have autobiographical significance and, amongst the clues given to the readers that the hero figure was based on Ernest himself, is the intimation that Astria was based on none other than his own "beautiful wife".

There are some remarkable science-fiction predictions in the book, such as radium as a major source of energy for societies. However, the fact that Ernest's name does not figure in analyses of science fiction writings going back hundreds of years and up to relatively recent predictions (Aldiss 2001 and Wikipedia) may be a reflection that he never achieved fame as an author with just two published books, or that at the time the book was published it was not seen as a work of science fiction. In fact, unlike his second book, no evidence of any publicity for this book has been found.

In summary ten significant predictions can be found in the book of which four could be said to have been successful predictions, while three may have been based on earlier predictions or existing technologies. This represents a success rate of 40%, which is impressive for the nature of the predictions and is easily on par with many well-known science fiction writers.

The 1911/12 Northern Exploration Company Prospectus and 1913 Marble Island Prospectus

Both of these NEC prospectuses were no doubt compiled with a large involvement by Mansfield, and the third picture in the 1913 prospectus is a studio portrait of Ernest Mansfield, which is the same picture that appears in the 1911/12 Prospectus, and is captioned "Early Pioneer". A studio portrait of the Rev. Gardner has the same caption. A large proportion of the book is taken up with the 83 full-page monochrome photographs taken at "Marble Island", and Ernest appears in many of the pictures.

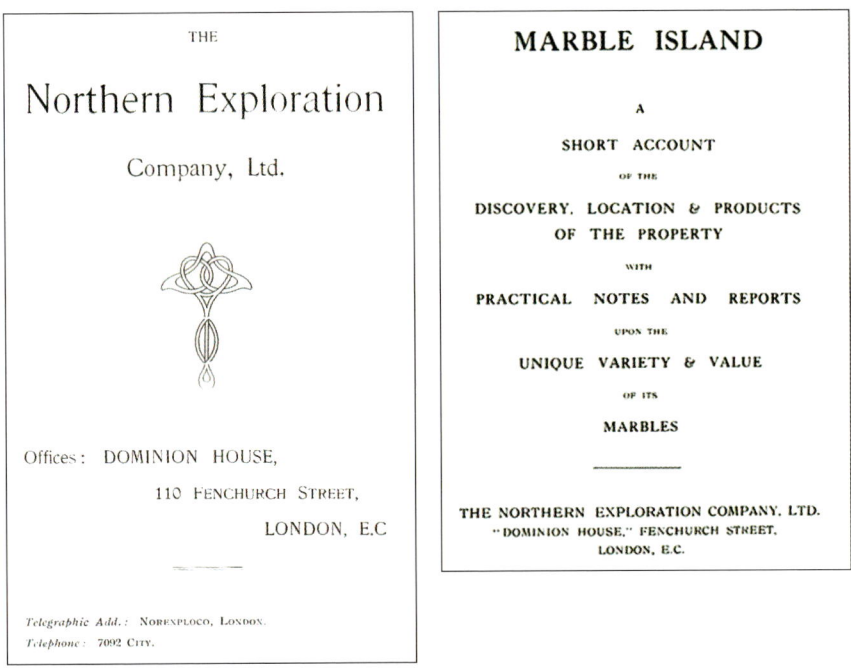

Left: The 1912 Prospectus title page. Courtesy of Richard Gardner
Right: The Marble Island Prospectus title page. Courtesy of Richard Gardner

Ralph Raymond

In the same year that the Marble Island book was produced, Ernest published his second semi-autobiographical novel entitled *Ralph Raymond* (Mansfield 1913). The story covers 344 pages and has eight drawings by the well-known illustrator of the day, Louis Gunnis.

In contrast to the apparent lack of promotional material associated with his first book, the second book appears to have been given plenty of publicity. The *Everyman* magazine (9.5.1913) presented a summary under the heading of "Book of the Week" and particularly emphasised that the book contained "Breathless excitement and dramatic incident". It ended by stating that "Mr. Mansfield, it may be presumed, is desirous of affording his public entertainment, and there can be no possible doubt that he has amply achieved this."

In June 1913 a long article, taking up most of a page in the newspaper, appeared in the *Wanganui Chronicle* (3.6.1913). It was a good summary of *Ralph Raymond* and must have been good publicity for the book. Much of it is taken directly from the book, and as it appeared in the *Wanganui Chronicle* it was probably written by Ernest. The first part reads thus:

> ERNEST MANSFIELD
> EXPLORER, PROSPECTOR, PROMOTER AND AUTHOR
> "RALPH RAYMOND"
> A STORY OF LOVE AND ADVENTURE
> A NEW BOOK OF LOCAL INTEREST
> A COMING DRAMA AND A PROMISED PICTURE PLAY
>
> Ernest Mansfield is a name still well remembered in Wanganui. Some years ago the owner of this name was a local resident, just one of the crowd, with seemingly no more alluring prospects than those of the ordinary man in the street. But he was a man fired with big ambitions, a man wanting more elbow room than this country could offer. In some respects he was a rough diamond, but those who knew him best, appreciated his extraordinary qualities and the force of his unique personality. The field of industrial occupation held no attraction for him. Commerce wooed him in vain, though he toyed with it long enough to prove himself a born hustler, capable of winning business while other men, older and more experienced, merely wore out their shoes. Gold was the magnet that drew him. In the quest of the precious metal lay the possibilities of big things. [...]
> And now from the pen of this man whose own life is evidence that truth is oft times stranger than fiction, there comes a book — a novel — which all lovers of the melodramatic will delight to read. "Ralph Raymond" (Stanley Paul and Co., London), may not be a gem of literary perfection, but it is at least a good yarn, and on its romantic side Mansfield has allowed his brilliant imagination to have full play. [...] The plot is very cleverly worked out and is rich in stirring situations and dramatic incidents, so that it is not surprising to learn that the book is already in the hands of the dramatist and that it is also to form the subject of a moving picture-play. Apart from the intrinsic merit of the story, as such, there is some good moralising in the book.

The semi-autobiographical style of the novel is established early in the story, although the overall murder-mystery plot happily seems too extreme to be founded on fact. The house called "The Laurels" "in a straggling village in a picturesque spot in East Anglia" is a play on Ernest's homes "The Limes", and the heroine's

RALPH RAYMOND BY ERNEST MANSFIELD to PROSPECTORS EVERYWHERE SECOND EDITION *First Published 1913* WITH EIGHT ILLUSTRATIONS BY LOUIS GUNNIS London STANLEY PAUL & CO 31 Essex Street, Strand, W.C. [*Dramatic and Translation rights reserved. Copyrighted in the United States of America by Ernest Mansfield.*]	*Wanganui Chronicle 1 Nov 1913* READ... **"RALPH RAYMOND,"** (Stanley Paul and Co., London) A **THRILLING ROMANCE,** By ERNEST MANSFIELD. Erstwhile of Wanganui and now a noted explorer. The book is well bound, and graphically illustrated by Louis Gunnis. A story of **Love and Adventure** throbbing with realism and crowded with dramatic incidents. On Sale at all Booksellers, 3s 6d.

Left: "Ralph Raymond" title page. Book in the current authors' possession
Right: Advertisement for "Ralph Raymond". Source: Wanganui Chronicle 1 November 1913

name Berice is a variant of Ernest's daughter Bernice. The book also includes many philosophical beliefs and opinions relating to gold prospecting and the people involved in the business within it, many given in the *Wanganui Chronicle* article. These would undoubtedly not have been appreciated by some of Ernest's many benefactors and associates had they taken the trouble read the book.

When Ralph Raymond is arrested, jailed and tried for murder, he makes this statement from the dock:

> Ralph Raymond, have you anything respecting this you would wish to say?
> "Only this, my lord," said Ralph, coming forward in the dock and gripping the rail to steel himself against succumbing to the mental and physical strain the long ordeal had imposed on him. "On behalf of those countless thousands who are trudging along the wilds, wildernesses, mountains, glaciers, valleys and streams, in all parts of the world, I thank you. I was one of them a short time ago. One of those despised men, who find the means of investing capital, and do the pioneer work that will pave the way for the erecting of new cities, railroads, and many other enterprises that follow in the wake of successful mining. This life makes one self-reliant, it never makes a coward! For your kindly reference to us, as a class, I thank you."

Later in the book it is stated: "Ralph is a fine man [...]. Why is it the people have such a contempt for his calling? Why should they look upon mining engineers as swindlers?". Miners and prospectors are further praised in the book as playing "one of the most important parts in the world's history, and the privations they have to endure sometimes must be terrible". Various New Zealand locations and goldfields are referred to that are real places: Goldsborough, Kokitiki, Kumara, Moonlight Creek, the Grey Valley and River, Teremakau and Westport. These are all on the South Island, near Greymouth. One of the NEC Prospectuses states that Ernest had worked at the Brunnerton mine near Greymouth.

Towards the end of the book, there are some extremely philosophical opinions relating to gold prospectors. The comments about "disgraced parsons" and "unworthy doctors" seem incredible considering the Rev. Gardner and Dr Salter probably funded Ernest's early Spitsbergen trips only a few years earlier:

> When a new alluvial goldfield is getting on in months there is always an influx of undesirable characters. Broken-down barristers, disgraced parsons, unworthy doctors, and disreputable scions of a wrecked nobility. These gentry having exhausted every means of existing amidst the civilisation they were nurtured in now infest the goldfields in the hope of obtaining a new lease of their miserable lives and, for a time, they thrive exceedingly well on their deferential sponging methods. They meet at the hotels to tell their glorious past, and the "what was" period they revelled in.

Comparisons between Astria – The Ice Maiden and Ralph Raymond

The biographical details in Ernest's first novel laid out in the first two chapters are quite obvious and appear factual, whereas in *Ralph Raymond* the biographical content is spread thinly throughout the book, so it is less evident and harder to separate from the fiction. *Ralph Raymond* also has many personal philosophical beliefs and opinions which are not present in *Astria – The Ice Maiden*. However, the latter has some interesting scientific predictions. *Astria – The Ice Maiden* covers the later Svalbard period of Ernest's life, whereas *Ralph Raymond* covers the earlier Australia and New Zealand period. There is only one very small mention of the British Columbia phase in *Ralph Raymond* and no mention of it in *Astria – The Ice Maiden*.

Both books are well written but inevitably are now dated in style. *Ralph Raymond* gives the impression of being more professionally produced, probably benefiting from Ernest's experience of the first book. Both convey something of the background and personality of the man and complement each other in that respect. Ernest clearly intended to record something about his life and his opinions, albeit in a disguised form. It is only now with the benefit of access to all the other material written about him, that we can fully appreciate the significance of these books. As he explained in the dedication at the beginning of *Astria – The Ice Maiden*, he has taken Dr Salter's advice that simply writing an autobiography "would smack of egotism".

Ernest's personality
– was he a dreamer or a swindler?

ERNEST MANSFIELD DIED IN a Leeds nursing home, now part of Leeds General Infirmary, on 1 December 1924 at the age of 64. It is not known why he was in the nursing home. His death certificate (1.12.1924) records the cause of death as kidney stones and blood clots following surgery. The address of the deceased and his widow Margaret Mansfield was given on the certificate as 96 Park Hill, Clapham Park, London. This is the same address as that on Bernice Zoe Mansfield's visiting card that appears inside the bound copy of *Astria – The Ice Maiden* in the possession of one of the current authors. No newspaper obituary has been found in the UK. However, there was a very short death notice in *The Times* (2.12.1924) and two obituaries in Norwegian newspapers from that period (see the chapter *Ernest in Svalbard*).

The authors have succeeded in tracking down only a few of Ernest's daughter Zoe's movements after his death in 1924 when she would have been 20 years old. In March 1928 her engagement was announced in *The Times* to a Ft Lt Cyril Ferdinand Briggs, with a wedding planned to take place in Bombay, India, on 18 May 1928. Ft Lt Briggs was a member of the R101 airship crew, which crashed on its maiden voyage to India in October 1930, killing 48 of the 54 people on board; however, neither Ft Lt Briggs nor Zoe were listed amongst the casualties or survivors, and the relatives of Ernest's wife Margaret (nee Booth) believe that Zoe did get married in India and later returned to London.

Despite the huge amount of material written about Ernest during his lifetime by himself and by others, he remains somewhat of an enigma and is still surrounded by several mysteries, anomalies, and contradictions. One hundred years on, it is difficult to comprehend what actually happened in some of these situations and hence piece together facets of his character that relate to them. From articles written by his contemporaries he was clearly popular with both business colleagues and those working for him and was clearly effective at making friends and establishing business partnerships in new situations as he travelled around the world.

His communication skills would not have been restricted to his role as a journalist and author. His verbal skills, evident from the number of speeches he is known to have made, would have meant that he was always good company and integrated well at any level of society. After Ernest's death Birger Jacobsen wrote: "Mansfield was of a magnificent character with a strongly developed poetic vein and an exaggerated imagination". Dr Salter wrote of him: "Talkative and given to exaggeration… His geese were always swans." There are also several accounts of his involvement with clubs and societies around the world.

He was extremely committed to his chosen career, dedicating most of his time to it. He would have spent large amounts of time travelling at a time when the only means was by sea, taking weeks, if not months, to move from one continent to another. This would have been the ideal time to spend writing. However, little time would have remained for family life and he seems to have had a lifestyle which today might be associated with a workaholic.

Notwithstanding the more positive elements of his personality, plainly not everything went his way, and not everyone he did business with was comfortable with the results. One can only wonder if some of the peculiar and irrational situations that arose in his career might be explained now with the benefit of hindsight and the resulting wealth of information about him that is accumulated here from many sources. Hence it is tempting now to try to analyse some of the events that happened and to seek answers to some of the uncertainties that have arisen while compiling his biography. There are considered below in chronological order:

Was Ernest a party to substituting the Lydia gold samples in New Zealand with the intention of misleading the investors?

At the time there were many newspaper reports and readers' letters about the mix-up and a false, and perhaps fraudulent, assay of gold samples from the Lydia mine. Ernest was in the forefront of the correspondence on the subject, robustly denying that he or his associate, Dr Hatherley had been involved with anything untoward. After several other apparently unproductive prospecting and promotional ventures over the next year, he suddenly left New Zealand ostensibly on a visit to the "old country". Unfortunately, the authors have found no conclusive evidence to resolve this question.

Was Charles Plowman related to Ernest?
Ernest's full name was Richard Plowman Ernest Mansfield and Plowman is both an unusual middle name and surname, so it is possible that Charles Plowman, Mansfield's associate in British Columbia, was his cousin on his mother's side. Not much is known about Charles Plowman and he did not appear on the electoral role in the region at the time. However, he was clearly one of Ernest's most loyal assistants in British Columbia and carried out many financial transactions on his behalf. Ernest expressed his "unbounded admiration for Mr. Plowman's rustling qualities".

Was Ernest's orchestrated incarceration in Nelson jail successful in getting more funds from the French investors, or was it a rather naïve publicity stunt that backfired?
Ernest certainly made the most of the publicity of being in jail and used his journalistic skills to write almost daily long newspaper articles that were published locally while he was in prison. He must have relied on the poor communication channels of the day to ensure that the investors on the other side of the Atlantic would not read his articles before the incident was over. As the result of the telegrams sent to France, the workers did get paid within a week or two, but Ernest destroyed his relationship with the syndicate's financial backers in the process. He initiated legal action against the syndicate, but as he did not record the outcome and the result was not chronicled elsewhere, it is most probable that he lost. However, he remained in the area for another four years and apparently worked successfully with other investors.

Why did Ernest move to England in 1904 and settle in the village of Goldhanger?
After travelling the world for ten years as a gold prospector and promoter of gold shares, it would seem irrational to move to the small village of Goldhanger where there has never been any gold! We know that there were distant family connections between Ernest and the Goldhanger Rector via the Timmins family, but that would not seem enough to justify him parting from his committed lifestyle. It now looks possible that the Rev. Gardner already had an insight into the potential mineral wealth of Svalbard before 1904 via his friend the Earl of Morton, and that he invited Ernest to come and join them. Perhaps this, and Ernest's desire for his new Scottish wife's child to be born in the UK and in one of the Rector's large houses, formed the basis of the move. It is just too much of a coincidence that he arrived in Goldhanger only to find the Rev. Gardner, the Rev. Hon. Bryon and Dr Salter, who were all wealthy and living within a three mile radius, keen to join with him in a new Svalbard adventure. It may have also been convenient to leave British Columbia at that time and not go back to either New Zealand or Australia.

Where was Ernest between mid-1901 and the move to Goldhanger in 1904?
This is one of the gaps in the biography that it has not been possible to fill with facts. The information that is available for these years is scanty and inconclusive. The newspapers in the Nelson area of British Columbia recount that he left B.C. in mid-1901 and that he founded a new company there in 1902, but no evidence of him returning to B.C. after 1901 has been found. His divorce was announced in 1902 in New Zealand, but presumably he did not have to be there in person. The article about Western Australia in the Wanganui Herald by a "Wanganui Boy" in 1903 was very likely written by Ernest, but again he was not necessarily there himself. A "Mr Mansfield" taking part in a play in Wanganui in 1903 was not likely to be Ernest unless other evidence could be found to confirm the fact. Finally, Dr Salter's Diary states that in 1903 Mansfield returned from New Zealand having been at the Wei Hei mines, but this is unsupported by the newspaper articles that usually followed his activities in New Zealand.

Another possibility is that he may have moved to the UK without leaving a trace of where he was situated. His wife was Scottish-born, he had an address in London at the end of his life and he died in Leeds. Where was his main UK connection and why London and Leeds? He could have gone from B.C. to Scotland or to London. Did he move to London because of his connection with Mr Timmins and through that find out that he should move on to Goldhanger to be near the two most interesting persons, the Rev. and the Dr? Perhaps London is our answer, but we do not know for sure.

Did Ernest have an incredible foresight in making the scientific predictions that are in his first book "Astria – the Ice Maiden", or are they just coincidences?
Very few copies of his first book seem to exist today. It was produced as a low cost paperback and, unlike his second book, does not appear to have been well promoted. So it is likely that sales were not high and few people read it. With no previous track record of authorship of novels, it would not have been known at the time if Ernest's predictions had merit. One hundred years later however, the predictions of a technological nature in the book, relating largely to his knowledge of the recently discovered minerals, seem impressive to the authors – who claim no expertise in the history of Science Fiction – and it could be stated that these might be more than just coincidences.

Was Ernest's membership of Freemasonry on three different continents used to befriend and engage wealthy investors in his prospecting schemes?

Ernest was known to be a Freemason while in the UK as his name is on a plaque identifying him as a Grand Master in the Easterford Lodge at Kelvedon in Essex, which is seven miles from Goldhanger and the lodge that Dr Salter was associated with for many years.

In the past, Freemasonry in the UK was surrounded in secrecy and assumed to be the province of wealthy businessmen and professionals. Amongst the symbolism involved, gold and marble play significant parts. These were the materials of special interest to Ernest. This is most probably coincidental however, as both are rare, expensive and associated with wealthy lifestyles.

We know Ernest befriended two prominent Freemasons, Dr Hatherley in Wanganui, New Zealand and Dr Salter in Essex, UK and both became enthusiastically involved in his prospecting and mining investments. No evidence has, however, been found to suggest that he was a member of a Masonic Lodge in the Nelson area of Canada. Freemasonry is not mentioned in Ernest's own writings, although at the time many committed masons would not have broken the secrecy vows. Dr Salter did not have this concern however, as he wrote extensively in his diaries published after his death of his own role in Essex Freemasonry. Since Freemasons are a brotherhood, committed to helping each other, it may be assumed that this helped to forge closer bonds between Ernest and other members, but regarding this as a decisive factor in the relationships would only be speculative.

Did Ernest know at an earlier stage that the Svalbard marble was unsaleable, and why did so many "experts" from both sides of the Atlantic give such glowing testimonials in the 1913 Marble Island Prospectus when it transpired that the marble was not commercially viable?

Ernest's brother-in-law David Booth recorded in his 1912/13 Day Book, which covered his assignment on Marble Island, that he wrote several letters to the company informing them that he was having great difficulty in finding samples of marble that did not crumble when removed from below the ground. One wonders why this information had no impact on NEC plans. David Booth was neither mentioned nor thanked in the glowing reports that Ernest wrote at the end of the season's work at the quarry, nor was he mentioned in the Marble Island Prospectus along with others who were praised for their efforts. Perhaps he had been too honest about the marble.

Two theories have been put forward as to why the marble crumbled: 1. The marble was originally highly fractured, but stuck together with water that had seeped into all the cracks and then frozen. When removed from permafrost conditions the ice unfroze and the marble crumbled, and 2. The marble at that particular location had been subject to ancient geological movements, causing it to have an inherent crazed structure and resulted in crumbling when machined. Both reasons in fact imply that the marble was in an unstable condition and either "glued" together by ice crystals or kept compact as long as it was confined in the ground. In both cases small selected pieces, such as the samples shown in the Marble Island Prospectus, may have remained intact, but larger blocks and slabs would have broken up. This could account for the glowing testimonials given by the experts in the UK, France and the USA, who probably only saw small samples.

Why did Ernest return to Australia in 1914, but apparently not go to New Zealand with which he had much greater links?

Ernest spent most of his early life, up to the age of 35, in New Zealand but had only spent a year in Western Australia in around 1897. Two possible reasons for going back to Kalgoorie and Lenora have been identified: To belatedly tidy up his late father's affairs, which included a short-lived involvement with the Harbour Lights and Marda mines, and also to promote his second book which has a significant Australian content. Several reviews about the book appeared in the Australian press at the time.

Did Ernest ever make a fortune, or did he manage to just miss out on several big opportunities?

In *Astria – The Ice Maiden* Ernest wrote: "I made a couple of fortunes and got through them", but he does not tell us at which location these fortunes were made. However, no evidence of these fortunes or the lifestyle that might have gone with them has been identified.

Fortunes were made by some in the Waikato district of New Zealand in the past where the Lydia and other Mansfield claims were located, including at Waihi where huge fortunes were made, but there is no evidence that Ernest struck lucky there other than by trading the shares. Reports from his 1897 Western Australia experience would seem to indicate a close association with the Hurst family, who did seem to make their fortune from gold mining in that region, but there is little indication that Ernest shared in their fortunes. During his 1914 visit to Kalgoorie and Lenora he had some involvement with the Marda mine, and much later the Marda area did produce vast profits, but again there is no sign that our man shared in them.

Ernest's stay in the Klondike was short and there is no suggestion that he made a fortune there. The five-year period he spent in the West Kootenay region of British Columbia is perhaps the most likely place where he could have done well, and there certainly were fortunes made and lost in West Kootenay. The city of Spokane, Washington, in particular owes a great deal to the wealth generated in the Kootenay mines. Although Ernest said he "made and lost" a couple of fortunes during his short time in B.C., while he was involved in considerable wheeling and dealing, there is not much to suggest that any of them were particularly lucrative. The mines in Camp Mansfield did not ship any ore as far as it has been possible to determine. However, Ernest's sudden departure from B.C. too would indicate that he perhaps made more from share dealings than from any mineral extractions.

In 1899 Ernest travelled to London to attend The Greater Britain Exhibition with a mining colleague, a Mr Timmins. If that person was one of the Timmins family, who only a few years later made huge discoveries in Ontario that became known as the Hollinger Gold Mines close to Timmins City, then Ernest could well have missed out on one of the best prospecting opportunities of his life.

Ernest spent more time and energy prospecting in Svalbard than anywhere else, and it is undoubtedly here that he left his biggest mark on history. However, neither gold nor marble has ever been extracted from the archipelago in significant quantities to make a fortune for anyone.

Without doubt Ernest and his team did find significant quantities of coal on the island, yet it appears that the NEC never shipped any in large amounts. However, the coal seams identified by Ernest in 1905–06 and shown on his claims map by Sveagruva in the Braganzavågen area could well have been one of his missed opportunities as this today is the most productive mine area in Svalbard.

Is it possible that the Rev. Gardner and Dr Salter were never aware of Ernest's strong opinions about people in high places expressed in his second book Ralph Raymond published in 1913.

In *Ralph Raymond* there are references to "disgraced parsons", "unworthy doctors" and "Parson, doctor, actress, lawyer, in fact every branch of art and the professions called into line to boost and boom the share". In addition: "When a new alluvial goldfield is getting on in months there is always an influx of undesirable characters. Broken-down barristers, disgraced parsons, unworthy doctors, and disreputable scions of a wrecked nobility". This seems quite incredible considering that the Rev. Gardner and Dr Salter appear to have funded Ernest's early Svalbard trips. It is even more surprising that they seemed to have remained friends after the book was published. Dr Salter wrote glowingly of Ernest in his *Spitsbergen Reminiscences*, which was well after the publication of *Ralph Raymond* as he also refers to Ernest's death. Perhaps the best explanation is that his two friends in Goldhanger were never given the chance to read the book.

In his second book Ernest referred to "swindlers" in relation to prospectors and promoters. From what we have learnt, did he fit his own description of a swindler?

In the English language "swindler" has definite negative connotations such as cheat, conman, shark, crook, and fraudster, and it was one of many strong and emotive words that Ernest chose to use in the second book. Having investigated what we can about some of the mysteries and contradictions of his life, we have perhaps come a little closer to answering this pivotal question.

The writings of his contemporaries indicate that at the time Ernest was not seen as a swindler. Far from it, he seems to have been admired and treated with respect by his workers and associates. He was also well received by the members of the upper echelons of the society that he chose to cultivate and depend on for financial backing. He was plainly a popular and respected person in his day, so could hardly have had an undesirable reputation.

Ernest's well-developed modus operandi was to "boost and boom the shares" and then sell and get out, at the optimum moment. This and his director's salary were probably his main sources of income. At the time this was considered to be legitimate business practice, it was widespread and the proceeds were counted as a perk for many company executives. However, today it would be seen as "insider trading", would be considered to be unethical and is illegal in many countries.

In contrast, there is no evidence that he actually made a fortune from prospecting anywhere around the world and, because he never gave up trying to find that pot of gold, "dreamer" is an appropriate description. Perhaps we are left with one final question …

What then were his most notable achievements?

After one hundred years the most noteworthy achievement of Ernest Mansfield's life is undoubtedly the image that he left us of himself and his fascinating lifestyle. We have come closer to this through his two semiautobiographical novels, the numerous newspaper articles written by him, the company prospectuses, photographs and maps. That image is of a well-liked and charismatic individual, a musician, composer, writer, author, poet and journalist. An adventurer, prospector, promoter and publicist. A maverick, a visionary and a romantic, maybe a likable rogue and certainly a dreamer. An enigma and an enduring character of intrigue, still able today to convey to us something of his world.

REFERENCES

Ernest in New Zealand (c. 1878–1897)
Grey River Argus, New Zealand, of 27 September 1911
The Labourers' Union Chronicle, November 1873
Mansfield, Ernest 1910: Astria the Ice Maiden. Lonsdale Press, London
Mansfield, Ernest 1913: Ralph Raymond. Stanley Paul & Co, London
New Zealand *Evening Post* of 8 December 1906
New Zealand *Observer* of 4.3.1893
New Zealand Star of 19 October 1889
New Zealand *Otago Witness* of May/June 1896
Silvertonian, British Columbia 6.1.1900
The *Times*, London, November 1910
The *Thames Star* of May 1898 and 31 July 1912
Wanganui Chronicle, New Zealand
Wanganui Founders Index
Wanganui Herald, New Zealand
Wellington Evening Post of November 1906

https://www.bdmonline.dia.govt.nz
http://www.draughtshistory.nl/polish.htm
http://www.nzhistory.net.nz/culture/immigration/home-away-from-home/summary
http://paperspast.natlib.govt.nz
Te Ara (Encyclopædia of New Zealand). http://www.teara.govt.nz/en/history-of-immigration/8
http://www.wanganui.com/home
Wikipedia about Wanganui

Ernest in Australia (1897 and 1914)

Salter 1933: *Dr Salter of Tolleshunt D'Arcy in the county of Essex: his diary and reminiscences from the year 1849 to the year 1932* written by J.O. Thompson and published by the J. Lane Company.

Mansfield 1910: *Astria - The Ice Maiden*, published by Lonsdale Press, London.

Mansfield 1913: *Ralph Raymond*, published by Stanley Paul & Co., London.

NEC 1911: *Northern Exploration Company Prospectus*, published by NEC, London.

Ernest in Canada (1898–1901)

B.C. Minister of Mines Report, 1901, p. 851

Beaton 1.1902: "A New Canadian Glacier," *Canadian Magazine*, January 1902, Welford W. Beaton, p. 213–19, viewed at http://tinyurl.com/y9qdkch

Blake, 1988: *Valley of the Ghosts*, Don Blake, 1988, p. 28

British Columbia Gazetteer and Directory of Mining Companies, 1900–01, p. 61, 79, 104

Cairnes, 1935: *Descriptions of Properties, Slocan Mining Camp, British Columbia*, C.E. Cairnes, 1935, p. 227

Carter and Doug Leighton, 1980: *Exploring the Southern Selkirks*, John Carter and Doug Leighton, 1980, p. 48

Henderson's British Columbia directory 1901

Jacobsen 16.1.1925: Birger Jacobsen's obituary of Mansfield, published in *Oslo Aftenavis*

Journal of the Northwest Mining Association, 1900, viewed via Google Books

Kemp, 5.1900: *BC Mining Record*, Randall H. Kemp, p. 177–78. "A fortunate lady mine operator"

Kootenay Mining Standard, 7.1899: p. 58, viewed at archive.org

London Gazette, 8.1.1901, p. 187; 23.4.1901, p. 2,087; 13.9.1901, p. 6,063; 2.6.1911, p. 4,222

Mansfield 1910: *Astria the Ice Maiden*, Ernest Mansfield, 1910, p. 15

McNeil, 29.12.2010: E-mail from Ross McNeil, 29 Dec 2010

Mines Report, 1910: British Columbia Minister of Mines Report, 1910, p. 97

Mines Report, 1933: British Columbia Minister of Mines Report, 1933, p. 210–11;

NEC Prospectus 1911/12: Northern Exploration Co. Prospectus, 1911/12, p. 8

Nelson (B.C.) *Daily Miner*, 9.4.1899: "They are quite satisfied"

Nelson Daily Miner, 18.6.1901: "To operate Joker"

Nelson Daily Miner, 19.10.1900: "Ernest Mansfield in jail"

Nelson Daily Miner, 20.10.1900: "Talks of the Joker"

Nelson Daily Miner, 22.10.1900: "Waiting"

Nelson Daily Miner, 24.10.1900: "Taken from a county jail"

Nelson Daily Miner, 25.10.1900: "Will attach the property"

Nelson Daily Miner, 26.10.1900, quoting from *The Ledge*

Nelson Daily Miner, 9.6.1901: "To arrive this week"

Nelson Daily News, 30.7.2009: Reprinting an item from 7.1909

Porter, 2006: *Historical Archaeology at an Industrial Townsite: Lille, Alberta*, Meghan Porter, 2006, p. 5–17 viewed at library2.usask.ca/theses/available/etd-04052006-194712/unrestricted/Thesis.pdf

Slocan (B.C.) *Drill*, 11.4.1900

Spokane Daily Chronicle, 19.1.1900: "Lively Camp Mansfield"

Spokane Daily Chronicle, 23.2.1900: "In Camp Mansfield"

Touchstones Nelson Shawn Lamb archives: Notes in business card file on Mansfield Manufacturing Co.

Stock certificate 11.4.1907: Kaslo-Slocan Mining and Financial Co., in the author's collection

Stock certificate 30.8.1911: Auction house description for a stock certificate from the West Kootenay Mining Co. Ltd. dated 30.8.1911, provided by Ed Mannings of Nelson on 27.11. 2010.

The Kootenaian (Kaslo, B.C.), 11.5.1899: "South Fork wagon road"

The Kootenaian, 25.5.1899: "South Fork wagon road"

The Kootenaian, 5–12.11.1899: "Camp Mansfield notes"

The Kootenaian, 1.6.1899, 8.6.1899

The Kootenaian, 7.12.1899 "South Fork wagon road"

The Kootenaian, 7.12.1899: "Three groups sold"

The Kootenaian, 3.8.1899

The Kootenaian, 22.2.1900: "The South Fork Road"

The Kootenaian, 1.3.1900: "Camp Mansfield notes"

The Kootenaian, 18.10.1900: "Mansfield talks"

The Kootenaian, 27.12.1900: "Lime: Will probably be shipped from Kaslo to smelter"

The Kootenaian, 7.2.1901: "Some rich creeks"

The Kootenaian, 14.3.1901: "The Fletcher Group bond"

The Kootenaian, 28.3.1901: "To England"

The Kootenaian, 20.6.1901: "Mansfield: Back from England – will resume work"

The Ledge (New Denver, B.C.), 9.11.1899: "Excelsior Gold Mining Company"

The Ledge, 14.12.1899: "A big mining deal"

The Ledge, 8.3.1900: "Ten Mile Road"

The Tribune (Nelson, B.C.), 2.12.1899: "Large sums of money paid out,"

The Tribune, 29.12.1899: "Mansfield makes a deal"

The Tribune, 20.10.1900: "Arrested at instance of miners"

The Tribune, 23.10.1900: "Still in durance"

The Tribune, 24.10.1900: "Will be released today"

Victoria (B.C.) *Daily Colonist*, 5.12.1899

Victoria Daily Colonist, 28.12.1899, quoting *The Tribune*

Victoria Daily Colonist, 29.6.1900, quoting *The Tribune*

Victoria Daily Colonist, 25.11.1900: "A suit for damages,"

Victoria Daily Colonist, 3.11.1901 and 18.5.1902

Wanganui (NZ) *Chronicle*, 4.8.1899

Wanganui Chronicle, 9.6.1899: "Mr. Ernest Mansfield in British Columbia"
Wanganui Chronicle, 22.12.1899, quoting the *Nelson Daily Miner*
Wanganui Chronicle, 22.12.1899: "Rapid development in Camp Mansfield"
Wanganui Chronicle, 23.1.1900: "Kaslo woman makes a big sale of mines"

Ernest in England (1904–1924)

Argus 27.9.1911: The New Zealand Grey River Argus, *In Search of Gold* available within www.paperspast.natlib.govt.nz.

Birth certificate: Bernice Zoe Mansfield, September 1904, Chelmsford Registry Office, UK.

Booth 1912: David Booth's photograph album courtesy of members of his family.

Thompson 1933: *Dr Salter of Tolleshunt D'Arcy in the county of Essex: his diary and reminiscences from the year 1849 to the year 1932* by J.O. Thompson.

Death certificate 1.12.1924: Ernest Richard Mansfield, Leeds Register Office.

GPM 1919: Goldhanger & Little Totham Parish magazines, Colchester General Library.

Kelly's 1899: Kelly's Trade Directory for 1899.

Key 2005: *Little Totham – The story of a small village*, written and published by Lorna Key.

Mann 1908: Charles Mann's photo album, courtesy of members of his family.

Mansfield 1910: *Astria – The Ice Maiden* by Ernest Mansfield, published by Lonsdale Press, London.

Mansfield 1913: *Ralph Raymond* by Ernest Mansfield, published by Stanley Paul & Co., London.

NEC 1911/12: Northern Exploration Company Prospectus courtesy of the Gardner family.

NEC 1913: NEC *Marble Island*, Prospectus courtesy of the Gardner family.

NEC 1926: Northern Exploration Company Balance Sheet for 1926, Companies House, London.

ODNB: *Oxford Dictionary of National Biography*, www.oxforddnb.com: Salter, Dr John Henry.

Reilly 2009: *Greetings from Spitsbergen*, by John T. Reilly, published by Tapir Academic Press.

Sales 6.5.1911: Sales agreement between E. Mansfield and NEC, Companies House, London.

Times 27.6.1919: Company Meetings – NEC Ordinary General Meeting.

Times 7.4.1920: *Northern Exploration Company – Possibilities of the Territory*.

Times 18.12.1920: Letter from Dr F. G. Gardner.

Times 2.12.1924: Death Notice – Ernest Mansfield.

Times 13.10.1936: Obituaries – Reverend Frederick Thomas Gardner.

Ernest in Svalbard (1904–1913)

Amundsen, Birger 1983: *Der Schröder-Stranz expedition – historien om en ulykkelig ekspedisjon til Svalbard i 1912*. In: Svalbardboka 1983–84:8–28, Ursus forlag, Tromsø.

Avango, Dag 2005: *Sveagruvan. Svensk gruvhantering mellan industri, diplomati och geovetenskap 1910–1934*. Jernkontorets Bergshistoriska Skriftserie 44, Stockholm.

Barr, Susan 1985: *Ernest Mansfield – drømmer, svindler, eventyrer, gentleman*. In: Svalbardboka 1985–86. Tromsø, s. 159–173.

Barr, Susan 2003: *Norway – A consistent polar nation?* Kolofon, Høvik. Reissued 2010, Fram Museum, Oslo.

Dole, Nathan Haskell 1922: *America in Spitsbergen. The Romance of an Arctic Coal-Mine*. Marshall Jones Company, Boston. Vol.II.

The Geographical Journal. Published by: Blackwell Publishing on behalf of The Royal Geographical Society (with the Institute of British Geographers).

Hjelle, Audun 1993: *Geology of Svalbard*. Norsk Polarinstitutt Polarhåndbok No.7. Oslo.

Hoel, Adolf 1966: *Svalbards historie 1596–1965*, Vol. I–III, Oslo 1966.

Jones, A.G.E.: Unpublished manuscript entitled *Spitsbergen 1918–1919* in S. Barr's possession. The sources quoted are mainly *The Times* 1918 and 1919 and *Outspan* 15 April 1932 – Wild died in South Africa in 1939. The *Outspan* was presumably a South African newspaper or magazine that published Wild's reminiscences.

Lønø, Odd: *Norske fangstmenns overvintringer*. In: Norsk polarinstitutt Meddelelser no. 102 and 105, Oslo 1991 for the period 1795–1905. Norsk polarklubb's Polarboken 1993–94 and 1997–98 for the period 1906–1920.

Norsk Polarinstitutt (NP) biography archives, Tromsø.

Nathorst, A.G. 1917: *Obituary of Hans L. Norberg* in Ymer 1917 H.4 pp.319–23.

Oxaas, Arthur 1955: *Svalbard var min verden*. Aschehoug, Oslo.

Place Names of Svalbard, 1942: Norsk polarinstitutt Skrifter nr. 80. Oslo.

NEC cabins in Svalbard

Askeladden: Cultural heritage database of the Directorate for Cultural Heritage, Oslo.

Avango, Dag et al. 2008: *Archaeological expedition on Spitsbergen*, Dag Avango et al, 2008. Available at http://www.let.rug.nl/arctic/lashipa_web/LASHIPA-5_report-2008.pdf

Booth 1912: David Booth's "Day Book" of 1912–13 and associated photos, courtesy of his family.

Douglas 2011: http://www.douglashistory.co.uk/history/Places/camp_morton.htm

Governor of Svalbard / SMS: Sysselmann på Svalbard / Riksantikvaren: Reports on heritage maintenance in Svalbard (archives)

Hoel, Adolf 1966: Svalbards historie 1596–1965.

Lønø, Odd 1998: Norske fangstmenns overvintringer [Norwegian trappers' winterings]. Del 5, 1910–1920. Polarboken 1997–1998 p.81–136. Norsk Polarklubb, Oslo.

Lønø, Odd 2002: Norske fangstmenns overvintringer [Norwegian trappers' winterings]. Del 7, 1925–1930. Polarboken 2001–2002 p.77–160. Norsk Polarklubb, Oslo.

Mann 1908: Charles Mann's photo album, courtesy of his family.

NEC 1911/12: *Northern Exploration Company Prospectus* courtesy of the Gardner family.

NEC 1913: NEC *Marble Island* Prospectus courtesy of the Gardner family.

Orvin, Anders K. 1939: *The settlements and huts of Svalbard*. Norsk polarinstitutt Meddelelser nr.49.
Rossnes, Gustav 1993: *Norsk overvintringsfangst på Svalbard 1895–1940* [Norwegian wintering trapping in Svalbard]. Norsk Polarinstitutt Meddelelser Nr.127, Oslo.
Vanity Fair 20.8.1913: A *Struggle with Death in the Arctic*, by Ernest Mansfield.

Other information has been gathered by S. Barr on field trips.

Ernest's Literary and musical endeavours
Aldiss 2001: *Trillion year spree: the history of science fiction*. Brian Aldiss & David Wingrove. Published by House of Stratus
Mansfield 1910: *Astria – The Ice Maiden* by Ernest Mansfield, published by Lonsdale Press, London
Mansfield 1913: *Ralph Raymond* by Ernest Mansfield, published by Stanley Paul & Co., London
NEC 1911/12: *Northern Exploration Company Prospectus*, courtesy of the Gardner family
NEC 1913: NEC *Marble Island* Prospectus, courtesy of the Gardner family